NEUROMETHODS ☐ 16

Molecular Neurobiological Techniques

NEUROMETHODS

Program Editors: Alan A. Boulton and Glen B. Baker

1. **General Neurochemical Techniques**
   Edited by *Alan A. Boulton and Glen B. Baker*, 1985
2. **Amines and Their Metabolites**
   Edited by *Alan A. Boulton, Glen B. Baker, and Judith M. Baker*, 1985
3. **Amino Acids**
   Edited by *Alan A. Boulton, Glen B. Baker, and James D. Wood*, 1985
4. **Receptor Binding Techniques**
   Edited by *Alan A. Boulton, Glen B. Baker, and Pavel D. Hrdina*, 1986
5. **Neurotransmitter Enzymes**
   Edited by *Alan A. Boulton, Glen B. Baker, and Peter H. Yu*, 1986
6. **Peptides**
   Edited by *Alan A. Boulton, Glen B. Baker, and Quentin Pittman*, 1987
7. **Lipids and Related Compounds**
   Edited by *Alan A. Boulton, Glen B. Baker, and Lloyd A. Horrocks*, 1988
8. **Imaging and Correlative Physicochemical Techniques**
   Edited by *Alan A. Boulton, Glen B. Baker, and Donald P. Boisvert*, 1988
9. **The Neuronal Microenvironment**
   Edited by *Alan A. Boulton, Glen B. Baker, and Wolfgang Walz*, 1988
10. **Analysis of Psychiatric Drugs**
    Edited by *Alan A. Boulton, Glen B. Baker, and Ronald T. Coutts*, 1988
11. **Carbohydrates and Energy Metabolism**
    Edited by *Alan A. Boulton, Glen B. Baker, and Roger F. Butterworth*, 1989
12. **Drugs as Tools in Neurotransmitter Research**
    Edited by *Alan A. Boulton, Glen B. Baker, and Augusto V. Juorio* 1989
13. **Psychopharmacology**
    Edited by *Alan A. Boulton, Glen B. Baker, and Andrew J. Greenshaw*, 1989
14. **Neurophysiology**
    Edited by *Alan A. Boulton, Glen B. Baker, and Case H. Vanderwolf*, 1990
15. **Neuropsychology**
    Edited by *Alan A. Boulton, Glen B. Baker, and Merril Hiscock*, 1990
16. **Molecular Neurobiological Techniques**
    Edited by *Alan A. Boulton, Glen B. Baker, and Anthony T. Campagnoni*, 1990

NEUROMETHODS

Program Editors: Alan A. Boulton and Glen B. Baker

NEUROMETHODS ☐ 16

Molecular Neurobiological Techniques

Edited by

Alan A. Boulton

University of Saskatchewan, Saskatoon, Canada

Glen B. Baker

University of Alberta, Edmonton, Canada

and

Anthony T. Campagnoni

University of California, Los Angeles, California

Humana Press • Clifton, New Jersey

Library of Congress Cataloging in Publication Data

Main entry under title:
Molecular neurobiological Techniques / edited by Alan A. Boulton, Glen B. Baker, and Anthony T. Campagnoni.
   p.   cm. — (Neuromethods ; 16)
   Includes bibliographies and index.
   ISBN 0-89603-140-3
   1. Molecular neurobiology—Methodology.   2. Nucleic acid hydridization.   3. DNA probes.   I. Boulton, A. A. (Alan A.)
II. Baker, Glen B., 1947-   .   III. Campagnoni, Anthony T.
IV. Series.
   [DNLM: 1. Genetics, Biochemical.   2. Neurobiology.   W1 NE337G v. 16 / ·WL 102 M7185]
QP356.2.M643 1989
599′.0188—dc19
DNLM/DLC
for Library of Congress   89-2053
                                                                                           CIP

© 1990 The Humana Press Inc.
Crescent Manor
PO Box 2148
Clifton, NJ 07015

All rights reserved

No part of this book may be reproduced, stored in a retrieval system, or transmitted in any form or by any means, electronic, mechanical, photocopying, microfilming, recording, or otherwise without written permission from the Publisher.

Printed in the United States of America

Preface to the Series

When the President of Humana Press first suggested that a series on methods in the neurosciences might be useful, one of us (AAB) was quite skeptical; only after discussions with GBB and some searching both of memory and library shelves did it seem that perhaps the publisher was right. Although some excellent methods books have recently appeared, notably in neuroanatomy, it is a fact that there is a dearth in this particular field, a fact attested to by the alacrity and enthusiasm with which most of the contributors to this series accepted our invitations and suggested additional topics and areas. After a somewhat hesitant start, essentially in the neurochemistry section, the series has grown and will encompass neurochemistry, neuropsychiatry, neurology, neuropathology, neurogenetics, neuroethology, molecular neurobiology, animal models of nervous disease, and no doubt many more "neuros." Although we have tried to include adequate methodological detail and in many cases detailed protocols, we have also tried to include wherever possible a short introductory review of the methods and/or related substances, comparisons with other methods, and the relationship of the substances being analyzed to neurological and psychiatric disorders. Recognizing our own limitations, we have invited a guest editor to join with us on most volumes in order to ensure complete coverage of the field. These editors will add their specialized knowledge and competencies. We anticipate that this series will fill a gap; we can only hope that it will be filled appropriately and with the right amount of expertise with respect to each method, substance or group of substances, and area treated.

Alan A. Boulton
Glen B. Baker

Preface

It goes without saying that the principles and techniques of molecular biology are having and will continue to have a major impact on investigations into nervous system structure and function. It is becoming increasingly apparent to neuroscientists in all subdisciplines that a working knowledge of the language, approaches, and techniques of molecular biology is indispensable for their work. For these reasons, the editors have decided to devote this volume of *Neuromethods* to the techniques of molecular biology and their application to neural systems. There currently exist a number of excellent reference technical manuals that describe molecular neurobiological techniques in great detail, and many of these are cited within the chapters included in this volume. It was not the intention of the editors or authors of this volume to duplicate these efforts. Rather, our intention was to present to the neuroscientist who is relatively unfamiliar with these methodologies an understanding of how specific techniques are used to approach major molecular neurobiological problems as well as a set of techniques that work in the laboratories of the individuals writing the chapters. In some cases, there are duplications of techniques—these have been retained to illustrate the range of variability of the technique and/or the flexibility of the method to study different types of problems. We hope that the chapters will provide the reader with an understanding of the methods and their applicability to neurobiological problems; and, perhaps, suggest new directions for the reader's research efforts.

Anthony T. Campagnoni

Contents

Preface to the Series v
Preface vii
List of Contributors xvii

ANALYSIS OF BRAIN mRNAs BY TRANSLATION IN VITRO
David R. Colman

    1. Introduction 1
    2. Choice of Translation Systems 1
    3. Selecting the Starting Material 2
    4. Wheat Germ Extract Preparation 3
    5. Amino-Acid Mixture 4
    6. Master (Energy) Mix 5
    7. Preparation of Dog Pancreas Microsomes 6
        7.1. Dog Pancreas Membrane Preparation 6
        7.2. Stock Solutions 6
        7.3. Solutions 7
        7.4. Procedure 7
    8. Outline of mRNA Preparation 8
    9. Translation 8
        9.1. Assembly of the Translation Mixture 8
   10. Processing of Translation Mixtures 9
       10.1. TCA Precipitation (for ^{35}S Labeled
             Polypeptides) 10
       10.2. TCA Precipitation (^{3}H Label) 10
   11. Immuneprecipitation 10
      References 12

PREPARATION OF cDNA LIBRARIES AND ISOLATION AND ANALYSIS OF SPECIFIC CLONES
Sammye Newman and Anthony T. Campagnoni

1. Introduction 13
2. Background 14
   2.1. First Strand Synthesis 14
   2.2. Second Strand Synthesis 15
   2.3. Ligation to Vector 16
   2.4. Introduction of Recombinant Molecules into a Bacterial Host 18
   2.5. Vector–Host Systems 18
   2.6. Protocols for Screening the Completed Library 21
   2.7. Analysis of Specific Clones 22
3. Methods 24
   3.1. Preparation of a cDNA Library in Lambda gt11 25
   3.2. Amplification and Screening of the Completed Library 35
   3.3. Characterization of Positive Clones 37
   References 43

PREPARATION AND USE OF SUBTRACTIVE cDNA HYBRIDIZATION PROBES FOR cDNA CLONING
Gabriel H. Travis, Robert J. Milner, and J. Gregor Sutcliffe

1. Introduction 49
2. The Arithmetic of mRNA 50
   2.1. Total RNA 50
   2.2. Polyadenylated RNA 50
   2.3. mRNA Complexity 51
3. General Principles of Cloning 52
   3.1. cDNA Libraries 53
   3.2. Clone Isolation Requires Information 54
4. Clones of Tissue-Specific mRNAs 58
   4.1. Brute-Force Isolation 58
   4.2. Plus-Minus Screening 58

5. Subtractive Hybridization 60
   5.1. Standard Procedure 60
   5.2. Phenol Emulsion-Enhanced DNA-Driven
        Subtractive Hybridization 62
6. Future Applications 75
   References 76

ANALYSIS OF BRAIN-SPECIFIC GENE PRODUCTS
Robert J. Milner and J. Gregor Sutcliffe

1. Introduction 79
2. Patterns of mRNA Expression 80
   2.1. Northern Blotting 80
   2.2. *In Situ* Hybridization 82
3. Interpretation of Nucleotide and Amino-Acid
   Sequences 83
   3.1. Definition of Open Reading Frames 84
   3.2. Consensus Sequences 85
4. Sequence Comparisons 92
   4.1. Searching the Databases 92
   4.2. Assessment of "Homology" 93
5. Preparation of Antibodies to Encoded Proteins 97
   5.1. Antibodies to Synthetic Peptides 98
   5.2. Antibodies to Expressed Proteins 103
6. Characterization of Encoded Proteins 105
   6.1. Criteria for Protein Identification 105
   6.2. Biochemical Analysis 106
   6.3. Immunocytochemical Analysis 109
7. Summary: From Structure to Function 110
   References 110

ISOLATION AND STRUCTURE DETERMINATION OF GENES
Wendy B. Macklin and Celia W. Campagnoni

1. Introduction 117
2. Generation of Genomic Libraries 119
   2.1. Preparation of High Mol Wt Genomic DNA 119
   2.2. Preparation of Genomic Library 120

3. Isolation of Genomic Clones 125
   3.1. Preparation of Probe 125
   3.2. Clone Purification 131
   3.3. Isolation of Phage DNA 136
4. Structure Determination of Genomic Clones 138
   4.1. Preliminary Mapping 138
   4.2. Southern Blotting 139
   4.3. Building a Complete Map 140
   4.4. Subcloning Segments of the Gene 140
   4.5. Sequencing Subcloned Genomic DNA 144
5. Investigation of Transcriptional and Regulatory
   Elements of Genomic Clones 145
   References 148

GENE-MAPPING TECHNIQUES
Robert S. Sparkes

1. Introduction 153
2. Technical Advances 155
3. Anatomy of the Genome 155
4. Methods Used for Gene Mapping 156
   4.1. Genetic Linkage 156
   4.2. Somatic-Cell Hybridization 157
   4.3. *In Situ* Hybridization 160
   References 162

METHODS FOR GENETIC LINKAGE ANALYSES
M. Anne Spence

1. Introduction 163
2. Basic Gene Linkage Analyses 163
3. After Linkage Is Detected 171
4. Summary 173
   References 174

LINEAGE ANALYSIS AND IMMORTALIZATION OF NEURAL CELLS VIA RETROVIRUS VECTORS
Constance Cepko

1. General Introduction 177
2. Introduction to the Virus 177
3. Adaptation of Viruses as Vectors 179
   3.1. General Vector Design Strategy 179
   3.2. Packaging Lines and Methods for Virus Production 180
   3.3. Gene Expression 192
4. Applications to Neurobiology 194
   4.1. Lineage Mapping 194
   4.2. Immortalization of Neural Cells 204
   References 211

TRANSGENIC MICE IN NEUROBIOLOGICAL RESEARCH
Brian Popko, Carol Readhead, Jessica Dausman, and Leroy Hood

1. Introduction 221
2. The Production of Transgenic Mice 221
   2.1. The Isolation of Fertilized Eggs from the Ampulla of the Oviduct 222
   2.2. The Injection of DNA into the Pronuclei of Fertilized Eggs 223
   2.3. The Transfer of Injected Embryos into the Oviduct of Pseudopregnant Females 225
   2.4. The Analysis of Pups 227
   2.5. The Management of Data by Computer 228
3. Other Approaches of Introducing Foreign DNA into the Germline of Mice 229
   3.1. The Generation of Transgenic Mice Through the Use of Pluripotent Stem Cells 229
   3.2. The Use of Retroviruses to Introduce Foreign Genetic Material into the Germline of Mice 230

4. The Application of the Technique of Generating Transgenic Mice to Questions of Neurobiological Interest 230
  4.1. The Elucidation of Factors Involved in Regulating Gene Expression 230
  4.2. The Ablation of Specific Cell Lineages 231
  4.3. The Transformation of Specific Cell Types In Vivo 231
  4.4. The Generation of Mouse Mutants of Defined Origin 232
References 235

IN SITU HYBRIDIZATION
Michael C. Wilson and Gerald A. Higgins

1. Introduction 239
2. Applications of In Situ Hybridization to Problems in Neurobiology 240
3. Kinetics of In Situ Hybridization 245
  3.1. Stability of Hybrid Duplexes 245
  3.2. Stability of Hybrids Formed In Situ 246
  3.3. Dextran Sulfate 247
  3.4. Rate of Hybridization 247
4. An Overview of In Situ Hybridization Methods 248
  4.1. Preparation of the Tissue 248
  4.2. Pretreatment 250
  4.3. Probes for In Situ Hybridization 252
5. Controls for In Situ Hybridization 259
  5.1. RNase Pretreatment 260
  5.2. Heterologous Probes 260
  5.3. Other Controls of Hybridization Specificity 260
6. Simultaneous Detection of Multiple-Gene Products 262
  6.1. Immunocytochemistry and In Situ Hybridization 262
  6.2. Double-Label Hybridization 263

7. Quantitation of *In Situ* Hybridization: Problems and Potential 263
   7.1. Densometric Analysis of X-Ray Film Images 264
   7.2. Grain Counts at the Cellular Level 267
8. A Procedure for *In Situ* Hybridization 268
   8.1. Preparation of Tissue 268
   8.2. Preparation of ssRNA Probes 270
   8.3. Prehybridization Treatments 273
   8.4. Prehybridization 274
   8.5. Posthybridization Treatments 276
   8.6. Autoradiography 277
   References 278

Index **285**

Contributors

GLEN B. BAKER • *Neurochemical Research Unit, Department of Psychiatry, University of Alberta, Edmonton, Alberta, Canada*

ALAN A. BOULTON • *Neuropsychiatric Research Unit, University of Saskatchewan, Saskatoon, Saskatchewan, Canada*

CELIA W. CAMPAGNONI • *Mental Retardation Research Center, Department of Psychiatry and Biobehavioral Sciences, Neuropsychiatric Institute, Los Angeles, California*

ANTHONY T. CAMPAGNONI • *Department of Psychiatry, University of California, Los Angeles, California*

CONSTANCE CEPKO • *Department of Genetics, Harvard Medical School, Boston, Massachusetts*

DAVID R. COLMAN • *Departments of Anatomy and Cell Biology, and Pathology, The Center for Neurobiology and Behavior, Columbia University, New York, New York*

JESSICA DAUSMAN • *Division of Biology, California Institute of Technology, Pasadena, California*

GERALD A. HIGGINS • *Department of Neurobiology and Anatomy, University of Rochester Medical Center, Rochester, New York*

LEROY HOOD • *Division of Biology, California Institute of Technology, Pasadena, California*

WENDY B. MACKLIN • *Mental Retardation Center, Department of Psychiatry and Biobehavioral Sciences, Neuropsychiatric Institute, Los Angeles, California*

ROBERT J. MILNER • *Division of Preclinical Neuroscience and Endocrinology, Research Institute of Scripps Clinic, La Jolla, California*

SAMMYE NEWMAN • *Department of Psychiatry, University of California, Los Angeles, California*

BRIAN POPKO • *Biological Sciences Research Center, University of North Carolina, Chapel Hill, North Carolina*

CAROL READHEAD • *Division of Biology, California Institute of Technology, Pasadena, California*

ROBERT S. SPARKES • *Division of Medical Genetics, Department of Medicine, UCLA Center for Health Sciences, Los Angeles, California*

M. ANNE SPENCE • *Division of Medical Genetics, Departments of Psychiatry and Biomathematics, Neuropsychiatric Institute, Los Angeles, California*

J. GREGOR SUTCLIFFE • *Department of Molecular Biology, Research Institute of Scripps Clinic, La Jolla, California*

GABRIEL H. TRAVIS • *Department of Molecular Biology, Research Institute of Scripps Clinic, La Jolla, California*

MICHAEL C. WILSON • *Department of Molecular Biology, Research Institute of Scripps Clinic, La Jolla, California*

Analysis of Brain mRNAs by Translation in Vitro

David R. Colman

1. Introduction

In vitro translation systems are extremely important tools for studying protein biosynthesis. These systems have been most useful in revealing the intracellular sites of synthesis of numerous proteins, the nature of cotranslational proteolytic cleavage events, the process of core glycosylation, and the mechanisms underlying the interaction of nascent polypeptides with organelles (such as mitochondria) and with membrane vesicles (rough microsomes, RM) derived from the rough endoplasmic reticulum (RER) (*see* Blobel, 1980; Sabatini et al., 1982, for reviews). When programmed with the total mRNA from an organ, translation systems will synthesize virtually all the encoded polypeptides, and so can also be useful in identifying antigenically-related proteins. Although precipitating antibodies to the proteins under study have in the past been required for these analyses, it is now possible to synthesize from cloned cDNAs large quantities of individual mRNAs, that, when used to program an in vitro translation system, yield a single polypeptide that can be directly studied.

This chapter contains well tested protocols for the preparation of wheat germ extracts with a very low background and exceptionally high translational capacity when programmed with exogenous mRNAs. Since the preparation of reliable translation mixtures, like good cell-fractionation, is as much art as science, I have included as many hints from personal experience as possible.

2. Choice of Translation Systems

It is the author's experience that the wheat germ system offers the greatest overall advantage for general work. It is extremely low in cost (a 1 lb bag of wheat germ, at about $2/lb, will render up enough extract to last for *years*). Unlike the more commonly used reticulocyte lysate system, no nuclease treatment is necessary, and there is complete uniformity between extracts prepared from the same raw wheat germ package. The wheat germ system accepts dog pancreas microsomes and gives beautifully clear backgrounds in autoradiograms of in vitro synthesized labeled polypeptides, without the distortion of low molecular products that is sometimes encountered with the reticulocyte system, owing to the presence of large amounts of globin. The single disadvantage of the wheat germ system is that some high molecular proteins (in our hands, primary translation products greater than 150 Kd) are not translated efficiently, and for these polypeptides, any of the commercial reticulocyte lysate preparations are usually adequate. An excellent review and guide to the preparation of reticulocyte lysates from phenylhydrazine-treated rabbits has been published (Clemens, 1984).

For the vast majority of work, however, it should be stressed that the wheat germ system is most reliable. Commercial preparations of these extracts are also available, but we prefer the following homemade extract.

3. Selecting the Starting Material

It has been our experience that nitrogen-flushed, refrigerated, raw wheat germ obtained in 1 lb bags from health food stores give the best results. A 1 lb bag should be good for approximately 2 *liters* of complete translation mixture, so it is worth the effort to select a good batch of wheat germ. Buy one bag from each of several stores. Do not buy the open-bin, room-temperature-stored stuff, since this is sure to be inactive. Back in the lab, open each bag, and sniff. There should not be the faintest hint of rancidity. If there is, do not use that batch. When crushed in a mortar and pestle with a little water, the germ should smell like fresh-cut grass. When you identify a good batch by this method—it is usually that simple—rapidly aliquot the germ into 4 g aliquots (approximately), and freeze ($-80°C$). One aliquot will be used to prepare 6–7 mL of extract, enough for hundreds of 25 µL translations.

4. Wheat Germ Extract Preparation

(*See* Roberts and Paterson, 1973; Marcu and Dudock, 1974; Roman et al., 1978.)

The Day Before
1. Preswell about 40 g G25 (medium) Sephadex in 1 L of *column buffer:* 20 mM Hepes, pH 7.4; 50 mM KCl; 1.0 mM MgCl$_2$. This takes about 2 h. It is most convenient to do this in a 2 L Ehrlenmeyer flask. When swollen, stir the Sephadex and allow the heavy beads to settle. Pour off the "fines" in the turbid supernatant. Repeat with 1 L of fresh buffer.
2. Add 500 mL of fresh buffer to the Sephadex and autoclave the mixture. Cool to 4°C, add DTT to 2 mM, and pack a 2.5 cm wide column with about 160 mL of swollen Sephadex. This column should be set up in a cold room. Adjust the flowrate to about 75 mL/h and run through about 150 mL of *column buffer* with DTT (4°C). Close off column.
3. Wash out a mortar and pestle with 1% liquinox and abundant distilled water. The mortar and pestle should be absolutely clean and dried thoroughly. Wrap in aluminum foil and autoclave. Cool to room temperature and store overnight in a freezer, at –80°C (or less).

On Preparation Day
4. Remove mortar and pestle from freezer. In the cold room, strip off foil and pour liquid nitrogen into mortar with pestle in it. When all nitrogen has evaporated, add 3 g of wheat germ (stored at –80°C until last minute) to mortar and about 5 mL of liquid N$_2$ and begin to grind with a circular motion. (Wear thick gloves for this; the pestle is very cold.) Spread germ up over sides of mortar and use moderate pressure to produce a fine, white powder, almost as fine as talcum powder. As the N$_2$ evaporates, add more, so as to keep the inside of the mortar as cold as possible. Note: Too much liquid N$_2$ will make grinding very difficult; use just enough to cover the germ.

5. When grinding is complete, add powdered germ to 10 mL *extraction medium* in a 15 mL Corex tube: 1 mM Mg acetate; 90 mM KCl; 2 mM DTT. The mortar can be scraped clean with a Teflon policeman. Do not use metal.
6. Let the germ in the *extraction medium* stand on ice for 10 min. Centrifuge for 10 min at 10,000 rpm in an HB-4 (swinging bucket) Sorvall rotor.
7. Decant the supernatant from *under* the lipid layer with a Pasteur pipet. Discard the pellet.
8. Add to the supernatant Hepes to 20 mM and Mg acetate to 2 mM.
9. Spin for 10 min at 10,000 rpm at 4°C.
10. Adjust column flowrate again to ~75 mL/h. Let the buffer run down so that the surface of the Sephadex becomes "velvety" in appearance. Load 10 mL of the supernatant on the Sephadex column; let this enter, then add 10 mL *column buffer;* when the buffer has entered the Sephadex, carefully fill column (50 mL maximum) and keep a "pressure head" of 50 mL over the Sephadex. The rapidly moving tan-colored front is the active extract. This should move as a "bolus" within the column. The slowly moving yellow fraction is not collected.
11. Discard the initial 1 mL of the turbid front; collect the next 6–7 mL of the turbid fraction. Mix briefly.
12. Prepare 100 μL aliquots in small Eppendorf tubes, and drop into liquid nitrogen. When frozen, remove and store at –80°C (at least).

5. Amino-Acid Mixture

Prepare the following components.

alanine (ala)	100 mM
arginine (arg)	100 mM
asparagine (asn)	100 mM
aspartic acid (asp)	100 mM
cysteine (cys)	100 mM
glutamine (gln)	100 mM
glutamic acid (glu)	100 mM
glycine (gly)	100 mM

histidine (his)	100 mM
isoleucine (ile)	100 mM
leucine (leu)	100 mM
lysine (lys)	100 mM
methionine (met)	100 mM
phenylalanine (phe)	100 mM
proline (pro)	100 mM
serine (ser)	100 mM
threonine (thr)	100 mM
valine (val)	100 mM
tryptophan (trp)	20 mM
tyrosine (tyr)	2 mM

Each amino acid should be weighed out, put into H_2O and neutralized (check on pH paper) as it dissolves with a 1 M KOH solution. This is necessary for solubilization. Note that for tryptophan and tyrosine one cannot prepare 100 mM stocks. Add 1 mM DTT (final) to cysteine. I usually make up 1 mL of each amino acid, filter sterilize (0.22 μM filter) and freeze (–80°C). To make up amino acid mixtures, thaw each tube, warm to 50°C to make sure they are completely dissolved, and add 500 μL *tyr*, 50 μL *trp*, and 10 μL of each of the other amino acids. *Leave out* the one(s) for which you will be providing a labeled isotope. (In most cases, this will be methionine, which is available with the high specific activity ^{35}S moiety.) Make the final volume up to 1 mL.

6. Master (Energy) Mix

Master Mix

	μL
Amino acids, minus the one(s) to be labeled*	16
1 M HEPES (pH. 7.6), adjust pH with KOH	16
2 M K acetate	44.8
0.1 M DTT	1.6
10 mM Spermine	6.4
100 mM ATP	8
12 mM GTP	8
800 mM Creatine phosphate	8
Creatine phosphokinase, 5 mg/mL	19.2

*See section 5.

Stocks of the individual components may be made up, aliquoted, and frozen at –80°C. All solutions should be made up with very pure (18 mega Ohm) water that has been filtered (0.22 μM filter). In addition, it is a good idea to filter-sterilize the amino-acid mix, HEPES, and K acetate solutions. The ATP, GTP, creatine phosphate, and phosphokinase solutions should be made up in 10 mM HEPES, pH 7.6. All components should be mixed together *on ice*, divided into 40 μL aliquots, and frozen in liquid nitrogen.

7. Preparation of Dog Pancreas Microsomes

(*See* Scheele et al., 1978; Shields and Blobel, 1978; Walter et al., 1981).

7.1. Dog Pancreas Membrane Preparation

In order to analyze translation products that are normally synthesized in vivo on membrane-bound polysomes, it is frequently important to assess the capacity of the in vitro synthesized, nascent polypeptides to associate with rough microsomes. For this purpose, rough microsomes derived from dog pancreas have proven most useful. These membranes are particularly low in RNAse and are highly active in processing of polypeptide in vitro. A good batch of membranes carries with it sufficient signal recognition particle (SRP) to mediate protein translocation. The membranes are commercially available from several sources or can be prepared from a freshly obtained pancreas. In the latter case, the pancreas can be removed on the day of surgery from dogs that are routinely used for research purposes by cardiovascular surgeons at all major medical centers.

7.2. Stock Solutions

1. 1 M TEA (Triethanolamine) (adjust to pH 7.5 with acetic acid);
2. 2.5 M sucrose;
3. 4 M K acetate (adjust to pH 7.5 with acetic acid);
4. 1 M Mg acetate;
5. 200 mM EDTA;

6. 1 M DTT (dithiothreitol);
7. 100 mM PMSF (phenylmethylsulfonyl fluoride) (in ethanol).

7.3. Solutions

1. Buffer A: 250 mM sucrose/50 mM potassium acetate/6 mM Mg acetate/1 mM EDTA/1 mM DTT/0.5 mM PMSF/50 mM TEA (pH 7.5);
2. Buffer B: 250 mM sucrose/1 mM DTT/50 mM TEA (pH 7.5);
3. Buffer C: 0.5 mM Mg acetate/1 mM DTT/50 mM TEA (pH 7.5);
4. Sucrose cushion buffer: 1.3 M sucrose in Buffer A;
5. Suspension buffer: 250 mM sucrose/1 mM DTT/50 mM Tris (pH 7.5).

7.4. Procedure

1. Weigh fresh dog pancreas and take this weight as the original *volume;*
2. Scrape the pancreas to get rid of connective tissue;
3. Mince with a razor blade;
4. Tissue press into 4 volumes of *buffer A*. Homogenize (6 strokes) with a motor driven Potter-Elvejhem homogenizer #65;
5. Spin homogenate in a Sorval HB4 rotor (30 mL Corex tubes) at 2,500 rpm (1,000 xg) for 10 min. Collect supernatant into another set of Corex tubes and spin in HB4 rotor at 7,500 rpm (10,000 g) for 10 min;
6. Collect supernatant into Beckman Ti60 high speed tubes (24 mL capacity); underlay with 10 mL of *sucrose cushion buffer*. Spin in a Ti60 rotor at 45,000 rpm (140,000 g), 4°C for 2.5 h;
7. Pellets are collected and resuspended in about 15 mL of *buffer B* in a Dounce homogenizer. Dilute by adding 9 vol of *buffer C*;
8. Spin in Ti60 at 30,000 rpm, 4°C for 30 min;
9. Microsomal pellets are resuspended at ~80–100 OD$_{280}$/mL with *suspension buffer* and aliquoted (25 µl) and stored at –70°C.

8. Outline of mRNA Preparation

(*See* Liu et al., (1979)
For preparation of "free" and "bound" polysomal fractions from which total RNA can be extracted, see Colman et al. (1983). We prefer to use the following protocol for mRNA preparation from total RNA.

A maximum of 3 g of brain are homogenized in 25 mL (at room temperature) of a solution containing

6 M guanidinium *HCl* (ultrapure)
25 mM Na acetate (filter sterilized pH = 6)
10 mM DTT (ultrapure)
0.5% Triton (RNase/DNase-free)

The homogenate is layered over 6 mL of filter-sterilized 5.7 M CsCl, 100 mM EDTA (pH 6), in a Beckman SW27 tube, and spun for 20 h at 24,000 rpm (23°C). Carefully decant tube contents, invert, and let tube stand to drain completely. The total RNA pellet is taken up in 1.0 mL (60°C) sterile H_2O, extracted once with phenol:chloroform (1:1; equilibrated to pH 7.2 with 1 M Tris HCl buffer), 2x with chloroform, and precipitated (at −20°C) in Eppendorf tubes at least twice with 200 mM Na acetate (pH 6) and 2.5 vol EtOH. The RNA is pelleted, and passed (2x) over an oligo dT cellulose (type 7, Pharmacia) column in an appropriate solution (such as 0.5% SDS and 500 mM NaCl), eluted with H_2O, and precipitated (2x) with 200 mM Na acetate and EtOH. Precipitated, pelleted mRNA is washed with 80% EtOH, pelleted, *lightly* dessicated, and solubilized in H_2O at a concentration of about 0.5–1 µg/µL.

9. Translation

9.1. Assembly of the Translation Mixture

All components should be added on ice, and the last component to be added should be the RNA. For each RNA, whether total or mRNA, a titration curve should be determined in order to find the RNA concentration that gives the greatest stimulation, as assessed by TCA-precipitable counts. After an optimal RNA concentration is found, an autoradiogram of the in vitro synthesized polypeptide pattern should be prepared. High molecular weight products should be apparent, although, as with all reinitiating translation systems, these are never as abundant as those of low molecular weight. It may also be advisable to run a Mg^{2+} curve as well, since different mRNAs have different Mg^{2+} optima.

If membranes are to be added to the translation mixtures, add 0.1 µL (from a 80–100 OD_{280}/mL stock) per 25 µL translation. Membranes will reduce the incorporation of the mix by roughly 50%, so do not be surprised if the TCA-precipitable counts drop accordingly.

Translations should be carried out in a water bath set at 25°C for a period of 3 h. Control translation mixes should contain no exogenous RNA ("the blank"), and/or no dog pancreas membranes. The "blank" is an especially important control, since some batches of wheat germ have an unacceptably high background. A good batch should give *at least* a 20-fold stimulation (TCA-precipitable counts) over the blank when programmed with an optimal amount of RNA.

If dog pancreas membranes are used, it is important to *gently* vortex the translation mixture every 15 min or so, to keep the membranes in suspension.

Translation

	µL
Water	3.5
10 mM Mg acetate	3
Master mix	4
^{35}S Methionine	2.5
Wheat germ extract	10
mRNA, 1 µg or total RNA, 10 µg	1
Total	25 µL

10. Processing of Translation Mixtures

After translation is complete, an assessment of the incorporation of labeled isotope into the in vitro synthesized polypeptides is an important step. This gives an accurate indication of the efficiency of translation of a given mRNA and allows the investigator to "standardize" each translation for immuneprecipitation and/or gel electrophoresis. TCA-precipitable counts are determined in the following way.

10.1. TCA Precipitation (for ^{35}S Labeled Polypeptides)

1. Spot 2 μL from each translation mix onto 3 mm (filter) paper circles (diameter = 2.4 cm). Allow to dry.
2. Place circles into a large volume (200 mL) of ice cold, 10% TCA that contains casamino acids 1 mg/100 mL. Keep on ice for 30 min.
3. Transfer circles to 5% TCA (with casamino acids) for 15 min.
4. Place circles in boiling 5% TCA, 10 min.
5. Rinse in cold 5% TCA, transfer circles to 100% ethanol to dehydrate.
6. Dry circles thoroughly under a heat lamp, place in a scintillation vial with a fluor and count (^3H + ^{14}C channels).

10.2. TCA Precipitation (^3H Label)

1. Add 5 μL of the translation mix to 1 mL of ice cold water. Add 0.5 mL 1 M NaOH and 50 μg casamino acids. Incubate at 37°C for 15 min.
2. Cool on ice, add 1 mL 25% TCA + 50 μg casamino acids. Let this mixture sit on ice for 30 min.
3. Vacuum filter onto a GFC filter (glass fiber). Dry the filter under a heat lamp, and place in a scintillation vial with a fluor and count.

11. Immuneprecipitation

(from Goldman and Blobel, 1978)

If membranes were added either co- or posttranslationally, they can be pelleted (3 min in an Eppendorf centrifuge) and immunoprecipitants prepared from the pellet and supernatant. There are also more extensive manipulations that can be performed on the membranes, such as high salt washing, endoglycosidase treatment, proteolysis, and detergent extraction, prior to immunoprecipitation. Data derived from these treatments can yield an accurate picture of the biosynthesis of a given polypeptide.

1. Make up Solution A:
   190 mM NaCl
   50 mM Tris HCl pH 7.4   Filter sterilize
   6 mM EDTA from pH 7 stock   with a
   2.5% Triton X-100   0.22 μM filter
   10 U/mL Trasylol

2. Swell 50 mg Protein A Agarose or Sepharose in an Eppendorf tube with 1 mL of Solution A. This takes about 2 h with gentle agitation. Spin 1 min in Eppendorf centrifuge and discard supernatant. Add fresh Solution A and shake; spin; discard supernatant and repeat. Store swollen beads in 2x bead volume Solution A with 0.2% sodium azide added.
3. Translation sample should be in an Eppendorf tube. Make the sample 2% in SDS from a 20% SDS stock. Puncture cap and suspend tube for 2 min in a gently boiling water bath.
4. Add 4x the translation volume of Solution A. Mix thoroughly. Transfer to new tube.
5. Add serum (10 μL). Generally, an overnight incubation at 4°C is best.
6. The next day, shake stored beads and add 3x volume of swollen beads desired. Thus, for 10 μL packed beads/tube (enough for 10 μL serum), remove 30 μL of swollen beads in Solution A.
7. Incubate at room temperature for 2 h while gently shaking the tubes. Beads should not be allowed to settle.
8. Spin down beads in Eppendorf centrifuge (30 s).
9. Carefully aspirate off supernate with a drawn out Pasteur pipette.
10. Add 1 mL of Solution A to beads and shake for 5 min.
11. Repeat steps 9–11 five times.
12. Spin beads down. Boil the beads in the gel sample buffer (containing at least 5% SDS and 100 mM DTT) for 1 min.
13. Load sample buffer plus beads onto gel.

References

Blobel G. (1980) Intracellular protein topogenesis. *Proc. Natl. Acad. Sci. USA* **77**, 1496–1500.

Clemens M. J. (1984) In: *Transcription and Translation: A practical approach,* Hames, B. D. and Higgins, S. J., eds., pp. 231–270. Oxford, IRL Press.

Colman D., Kreibich G., and Sabatini D. D. (1983) Biosynthesis of myelinspecific proteins. *Meth. Enzym.* **96**, 378–385.

Goldman B. M. and Blobel G. (1978) Biogenesis of peroxisomes: intracellular site of synthesis of catalase and uricase. *Proc. Natl. Acad. Sci. USA* **75**, 5066.

Liu C. P., Slate D., Gravel R., and Ruddle F. H. (1979) Biological detection of specific mRNA molecules by microinjection. *Proc. Natl. Acad. Sci. USA* **76**, 4503–4506.

Marcu K. and Dudock B. (1974) Characterization of a highly efficient protein synthesizing system derived from commercial wheat germ. *Nucl. Acids. Res.* **1**, 1385.

Roberts B. E. and Paterson B. M. (1973) Efficient translation of tobacco mosaic virus RNA and rabbit globlin 9s RNA in a cell-free system from commercial wheat germ. *Proc. Natl. Acad. Sci. USA* **70**, 2330–2334.

Roman R., Brooker J. D., Seal S. N., and Marcus A. (1978) Inhibition of the transition of a 40s ribosome-met tRNA complex to an 80s ribosome-met tRNA complex by 7 methylguanosine 5' phosphate. *Nature (Lond.)* **260**, 359–360.

Sabatini D. D., Kreibich G., Morimoto T., and Adesnik M. (1982) Mechanisms for the incorporation of proteins in membranes and organelles. *J. Cell Biol.* **92**, 1–22.

Scheele G., Dobberstein B., and Blobel G. (1978) Transfer of proteins across membranes. Biosynthesis *in vitro* of pretrypsinogen and trypsinogen by cell fractions of canine pancreas. *Eur. J. Biochem.* **82**, 593.

Shields D. and Blobel G. (1978) Efficient cleavage and segregation of nascent presecretory proteins in a reticulocyte lysate supplemented with microsomal membranes. *J. Biol. Chem.* **253**, 3753–3756.

Walter P. and Blobel G. (1980) Purification of a membrane associated protein complex required for protein translocation across the endoplasmic reticulum. *Proc. Natl. Acad. Sci. USA* **77**, 7112–7116.

Walter P., Ibrahimi I., and Blobel G. (1981) Translocation of proteins across the endoplasmic reticulum I. Signal recognition protein (SRP) binds to *in vitro* assembled polysomes synthesizing secretory protein. *J. Cell Biol.* **95**, 451–464.

Preparation of cDNA Libraries and Isolation and Analysis of Specific Clones

Sammye Newman and Anthony T. Campagnoni

1. Introduction

The development of molecular cloning techniques over the past two decades has generated a powerful arsenal of tools for the study of gene structure and function. It is now possible to synthesize complementary DNA (cDNA) copies of mRNAs from virtually any source and to propagate these cDNAs in vitro in the form of a cDNA library. Initially, cDNA libraries were designed specifically to generate probes for the identification of genomic DNA clones. However, the discovery that eukaryotic genes are often composed of coding regions interrupted by introns made it clear that the study of gene expression must include direct analysis of mRNAs and their corresponding cDNAs. Comparison of individual cDNAs to their parent genes and to related cDNAs has since revealed novel mechanisms for the processing of genetic information and has helped to define the primary genetic defects responsible for many disease states.

The procedures and techniques used in cDNA library preparation and analysis are now fairly routine in many laboratories throughout the world. By studying the basic components of the overall process and the logic behind their use, one can gain an understanding of the technical progression from mRNA isolation to analysis of individual cDNA clones, and this understanding can then serve as a foundation for the rational design of individual cloning projects. Because of the way in which the field of molecular biology evolved, a number of different routes exist for achieving each step in the cloning and screening process. A comprehensive presentation of all cDNA cloning methods is not intended here. Rather, the intent is to provide an appreciation of the technical questions that must be addressed before embarking on any cloning project and the general approaches that have been successfully used to prepare and screen cDNA libraries. In addition, detailed protocols that have been successfully employed in our laboratory will be presented.

2. Background

The primary goal of any cDNA cloning project is to synthesize a population of double-stranded DNA molecules complementary to a mRNA template, join these ds-cDNAs to a vector DNA molecule, and propagate the recombinant cDNA:vector in vitro. The ability to accomplish this goal stems directly from the discovery and characterization of a number of naturally occurring enzymes and the recognition of their usefulness as biological tools.

2.1. First Strand Synthesis

In 1970, two groups independently described an unusual DNA polymerase found in tumor viruses and noted its ability to synthesize DNA using an RNA template (Baltimore, 1970; Temin and Mizutani, 1970). The enzyme, termed reverse transcriptase, was subsequently used to prepare single-stranded cDNAs suitable for use as molecular probes (Verma et al., 1972; Kacian et al., 1972; Ross et al., 1972) and as templates for production of ds-cDNAs (Rougeon et al., 1975; Efstratiadis et al., 1976). Today, reverse transcription of a mRNA template is the first step in the preparation of any cDNA library. The synthesis of the first cDNA strand requires mRNA template, reverse transcriptase, a suitable primer, and deoxyribonucleoside triphosphates (dNTPs). The mRNA template, usually in the form of polyadenylated (poly (A)+) RNA, can be purified in a variety of ways (e.g., see Berger and Kimmel, 1987), but the necessity for a pure, intact mRNA preparation cannot be overemphasized, since all subsequent steps can be traced back to this origin. Reverse transcriptase is commercially available as a preparation derived from either mouse Moloney leukemia virus (MMLV) or avian myeloblastosis virus (AMV). In addition to the polymerase function, reverse transcriptase preparations also possess RNase H activity, a feature that can obviously be quite counterproductive during first-strand cDNA synthesis, especially since the RNase H preferentially attacks RNA in RNA:DNA hybrids (Leis et al., 1973). The removal by RNase H of the poly (A)+ tail from the template RNA : primer hybrid or partial degradation of the RNA during first strand synthesis to produce small random primer fragments can preclude effective synthesis of full-length cDNAs. Although the complete, selective inhibition of the RNase H activity of reverse transcriptase does not seem possible, the first-strand synthesis reaction can be optimized based on the criteria of maximizing the yield of full-length single-stranded cDNAs and minimizing the production of small fragments (Krug and Berger, 1987). The primer molecule in the reaction usually consists of a stretch of thymidylate residues that selectively hybridizes to the poly (A)+ tail of the template RNA. Since reverse transcriptase catalyzes the sequential addition of nucleotide residues to the 3'-OH groups of the growing cDNA molecule, the use of an oligo(dT) primer to initiate cDNA synthesis at the 3' end of the RNA template should increase the probability of synthesizing full-length copies. In some instances, the oligo(dT) is covalently attached to one end of a linearized vector molecule, which then serves as a vector/primer (Okayama and Berg, 1982). Random or specific oligonucleotides can also be used to prime the first-strand synthesis reaction from points within the RNA.

2.2. Second Strand Synthesis

The next step in the production of a cDNA library is second-strand cDNA synthesis, resulting in a population of ds-cDNAs. There are two basic methods for achieving this goal. In the first method, the formation of a hairpin structure in the first cDNA strand at what was the 5' end of the mRNA is exploited to prime the synthesis of the second strand by the Klenow fragment of DNA polymerase I and/or reverse transcriptase. The single-stranded loop at the end of the resulting double-stranded molecule is then selectively removed by digestion with S1 nuclease (Efstratiadis et al., 1976). A disadvantage of this approach is that the S1 nuclease treatment almost inevitably causes the removal of some nucleotides derived from the 5' end of the RNA template and can introduce nicks into the newly synthesized double-stranded molecules. The second method eliminates the requirement for nuclease digestion by priming second-strand synthesis with fragments of the original RNA molecule generated by nicking with the endonucleolytic enzyme *E. coli* RNase H, which is introduced into the reaction mixture after first-strand synthesis (Okayama and

Berg, 1982; Gubler and Hoffman, 1983). In this procedure, the first-strand synthesis reaction is optimized to prevent hairpin loop formation usually by inclusion of Actinomycin D. Since the second of the two procedures generally results in a greater yield of full-length ds-cDNAs, it has in practice replaced the original S1 nuclease-mediated method.

2.3. Ligation to Vector

Once the ds-cDNAs have been formed, they must be rendered "clonable," or suitable for insertion into a vector molecule. As with other steps in the preparation of cDNA libraries, this task can be accomplished using one of several different routes.

The enzymatic joining of DNA molecules via complementary single-stranded "tails" was first reported in the early seventies (Jackson et al., 1972; Lobban and Kaiser, 1973). The success of the initial experiments was based in large part on the earlier characterization of two classes of enzymes. The DNA ligases were found in 1967 to join DNA molecules annealed by complementary, single-stranded ends (Gefter et al., 1967; Olivera and Lehman, 1967; Weiss and Richardson, 1967; Zimmerman et al., 1967). In order to generate a suitable substrate for the DNA ligase from a double-stranded DNA, the activity of another enzyme, calf thymus terminal deoxynucleotidyl transferase (terminal transferase), was exploited. The terminal transferase was used for its ability to sequentially add nucleotide residues to the 3' termini of DNA molecules (Kato et al., 1967). Thus, a "tail" of oligo(dA) added to the ends of one DNA molecule could be ligated to a stretch of oligo(dT) constructed on the ends of a second DNA molecule (Jackson et al., 1972; Lobban and Kaiser, 1973). The DNA molecules in these experiments had to be treated with an exonuclease before addition of deoxyribonucleotides, since the terminal transferase required a single-stranded region adjacent to the 3'-OH terminus for activity under the conditions used. These conditions were later modified to eliminate this requirement and permit addition of homopolymer tails to DNA molecules with blunt ends (Roychoudhury et al., 1976).

An alternative to the tailing procedure was pioneered by Mertz and Davis (1972), who cut double-stranded DNA with the then novel restriction endonuclease EcoRI to generate overlapping ends that were complementary and, therefore, amenable to the action of DNA ligase. Moreover, the DNA fragments joined in this way could subsequently be separated easily by restriction endonuclease digestion. The cutting and joining of diverse DNA molecules via restriction endonuclease recognition sequences is now one of the most widely used techniques in molecular biology.

In a related experiment, Sgaramella (1972) used the DNA ligase isolated from phage T4-infected *E. coli* (T4 DNA ligase) to join ds-DNAs directly, without generating cohesive single-stranded ends before ligation. The use of T4 DNA ligase was based on the observation that the enzyme is capable of joining double-stranded DNAs in a process now known as "blunt-end" ligation. The techniques of blunt-end ligation and EcoRI-mediated ligation were then combined by Bahl and colleagues (1976), who added synthetic decadeoxyribonucleotide duplexes containing EcoRI recognition sites (EcoRI "linkers") to blunt-ended double-stranded DNAs and then cut the molecules with EcoRI and ligated them to one another via the resulting cohesive ends. Thus, DNA molecules no longer had to be fragmented into smaller pieces in order to create the ends necessary for ligation to an EcoRI site on another molecule. Methylation of internal EcoRI sites prior to addition of linkers is now typically used to prevent digestion of full-length cDNAs during preparation of cohesive ends (Wu et al., 1987). The blunt-end addition of synthetic linker molecules can be performed using a homologous population of linkers, as was done by Bahl and colleagues, or by using two different sets of linkers that contain distinct restriction enzyme recognition sites (Kurtz and Nicodemus, 1981). Moreover, the double-linker method can be modified in conjunction with the preparation of ds-cDNAs by adding the first set of linkers before S1 nuclease cleavage of the hairpin loop (*see above*, Section 2.2.) and the second set afterward so that the linkers are added sequentially to facilitate directional cloning of the DNA molecules (Helfman et al., 1983).

With the advent of the techniques described above, the groundwork was laid for joining DNA molecules from virtually any source. Today, ds-cDNA molecules are commonly prepared for insertion into a vector molecule by either homopolymeric tailing or addition of linker fragments and restriction endonuclease digestion to create specific cohesive ends.

2.4. Introduction of Recombinant Molecules into a Bacterial Host

Once constructed, the recombinant DNA:vector molecule must be introduced into a living cell where it can be propagated for analytical purposes. This goal can be accomplished by transformation, a term describing the uptake of DNA to produce a cell with a new trait, by transfection of bacterial cells with naked phage DNA to produce infectious phage particles within the host cell, or by infection of host cells by phage that have been packaged into infectious particles in vitro. The transformation and transfection methods most commonly used today are based on the work of Mandel and Higa (1970), who found that treatment of *E. coli* with calcium ions results in the entry of phage DNA into the cells and the production of infectious particles. A modification of this procedure was used to introduce autonomously replicating plasmids into *E. coli* (Cohen et al., 1972). Subsequent improvements to these basic protocols, such as the addition of a heat-shock treatment, have been introduced to increase the efficiencies of transformation and transfection (Hanahan, 1985). Encapsidation, or packaging, of lambda phage particles in vitro results in the recovery of recombinant DNA genomes with a frequency as much as 100-fold over that achieved with transformation or transfection of bacterial cells with plasmid or phage DNA (Murray, 1983). Methods for preparing packaging extracts have been published (Hohn and Murray, 1977), but commercially available packaging extracts that are generally of high quality are now commonly used.

2.5. Vector–Host Systems

The design of cDNA vector molecules was an outgrowth of the study of plasmid biology. Cohen and colleagues developed a small, self-replicating plasmid, pSC101, that expressed resistance to tetracycline, thus carrying a selectable marker when replicated in tetracycline-sensitive organisms (Cohen and Chang, 1973; Cohen et al., 1973; Chang and Cohen, 1974). pSC101 was subsequently shown to have a single EcoRI site in a nonessential region, rendering it perfectly suited to the ligation of inserts into EcoRI-generated overlapping ends. This plasmid was immediately recognized as a useful vector for the introduction of foreign DNA into *E. coli*, and its utilization as such was rapidly followed by development of a number of vector molecules, including plasmids such as pBR322 (Bolivar et al., 1977) derivatives of bacteriophage lambda (Murray and Murray, 1974; Thomas et al., 1974), and cosmids (Collins and Hohn, 1978). In more recent years, the principle improvements in vectors have included introduction of multiple cloning sites, or polylinkers, to facilitate the introduction of inserts into a variety of restriction enzyme sites (Messing et al., 1981), inclusion of a beta-galactosidase promoter and portions of the lac Z gene to permit both selection of recombinants based on a chromogenic reaction and expression of foreign DNA sequences in vitro (Young and Davis 1983a,b), and inclusion of promoters for bacteriophage SP6 and T7 RNA polymerases for synthesis of RNAs from cloned DNA sequences (Green et al., 1983; Melton et al., 1984). Moreover, a number of specialized vectors now exist for the propagation of cloned DNAs suitable for nucleic acid sequence analysis (Vieira and Messing, 1982; Messing, 1983).

It is important to realize that the design of any cloning project must include careful consideration of the host organism best suited for use with the particular vector one has chosen. In many cases a suitable host has been specifically designed for use with a particular vector to facilitate selection of organisms harboring recombinant cDNA:vector molecules. For example, the insertion of a foreign cDNA into the lac Z gene of the phage vector lambda gt11 results in inactivation of phage beta-galactosidase function detectable only on a lac Z- host background (Young and Davis, 1983a). In addition to facilitating selection of recombinants, the gene structure of specific host organisms has been exploited to increase the efficiency of other steps in the screening process, such as identification of specific clones based on expression of foreign antigens in a protease-deficient host (Young and Davis, 1983a,b). A remarkable number of vector–host systems are now available, and many more will certainly be developed to meet the demands of the evolving biological sciences. Suitable host organisms are normally provided with most commercially available vectors, but in many cases a cursory knowledge of host cell genetics and vector structure will provide flexibility in tailoring a cloning/screening project to specific needs.

In choosing an appropriate vector–host system for a given project, several basic questions must be considered. For example, if cloning into a particular restriction site is desired, the preferred vector should contain that site in a location that will facilitate selection of recombinants when propagated in an appropriate host strain (e.g., cloning into the PstI site of pBR322 will result in loss of the plasmid's ability to confer ampicillin resistance to an ampicillin-sensitive host). Many cloning methods now incorporate the blunt-end addition of linkers to full-length cDNAs to create synthetic restriction enzyme recognition sites suitable for cloning purposes, so that the choice of restriction site is dictated by the vector. It is important to remember that restriction endonucleases are the natural products of, and in fact were originally identified in, bacterial hosts. To prevent digestion of artificially introduced cDNA:vector molecules by the host's restriction system, restriction deficient (hsdR-) hosts are routinely chosen for growth and maintenance of cDNA libraries. If the mRNA in question is known to occur in the tissue source in very low abundance, a vector–host system capable of generating a library consisting of a large number of recombinants will be required (for a discussion of the statistical probabilities of detecting mRNA sequences of various abundance classes, see Jendrisak et al., 1987). Generally, packagable phage vectors are preferable if large numbers (>100,000) of recombinants are required since packaging of recombinant phage genomes in vitro increases the efficiency of recovery of recombinant phage by 10–100-fold over the calcium transfection methods (Murray, 1983). Plasmid vectors are suitable for preparation of libraries with a smaller number of individual recombinants.

Perhaps the most crucial factor determining the choice of a vector–host system is the method by which the library will be screened. The most commonly used screening methods are based on nucleic acid hybridization or antibody recognition of polypeptide sequences expressed in vitro. Almost any cDNA cloning system is appropriate for screening with nucleic acid probes, provided, of course, such a probe is available. The nucleic acid probe may exist in the form of a previously selected cDNA, a synthetic oligonucleotide designed to react with a known cDNA sequence, or a mixture of oligonucleotides that represents all possible codons for a known amino acid sequence. If the identity or primary structure of the protein or mRNA in question is not known, subtractive library preparation or plus-minus screening using nucleic acid probes prepared from RNAs from various tissue sources or from tissues in specified stages of induction or development may be required. If antibodies are to be used as probes, an expression vector such as lambda gt11 (Young and Davis, 1983a,b) or one of the pUC plasmids (Vieira and Messing, 1982; Helfman et al., 1983) must be used. It is sometimes possible to use antibody probes to screen an existing library prepared in a "nonexpression" vector such as pBR322 (Plaisancie et al., 1984). However, the chances of successful detection of antigens expressed by recombinant clones are greatly enhanced by using a vector–host system specifically designed for this application.

2.6. Protocols for Screening the Completed Library

Once a probe or population of probes has been chosen, the mechanics of screening the library are fairly straightforward. By using the technique of colony hybridization (Grunstein and Hogness, 1975; Hanahan and Meselson, 1980) or plaque hybridization (Benton and Davis, 1977) to screen with nucleic acid probes, as many as 10^5 individual library members can be screened on a single culture plate. Nitrocellulose replicas of the master plate are prepared by briefly blotting the filter to the surface of each plate (containing either plaques in top agar or colonies plated on nitrocellulose masters), binding the transferred DNA to the nitrocellulose replica, and allowing hybridization of the bound DNA with a labeled probe to occur. Since the preparation of the nitrocellulose replicas does not damage the parent colony or plaque, positive clones can be selected from the master plate or filter based on the hybridization signal on the replica, and then propagated in pure culture for further analysis.

Screening a phage or plasmid expression library using antibody probes also involves the production of replica filters, but in this case the filters are immersed in a solution of primary antibody directed against the protein of interest (Young and Davis, 1985; Helfman et al., 1985). Generally, polyvalent antibodies are most useful since they recognize multiple epitopes. The use of a monoclonal antibody may rely on recognition of an epitope that is absent from or structurally altered in polypeptides synthesized from recombinant clones, or that may be obscured by virtue of binding to the nitrocellulose membrane. The specificity of the antibody preparation must be ascertained, for example, by Western blot analysis, prior to its use as a probe. Furthermore, most polyvalent antibody preparations must be preadsorbed to remove IgG components that bind coliform proteins and cause "false positive" results. The primary antibody–antigen complex immobilized on the nitrocellulose filter can be detected by several methods, including binding of

^{125}I-labeled protein A, or reactivity with antiimmunoglobulins conjugated with horseradish peroxidase or alkaline phosphatase (Young and Davis, 1985; Helfman et al., 1985; Helfman and Hughes, 1987; Mierendorf et al., 1987). It is worth noting that it may be necessary to screen 5–10 times as many recombinants using antibody probes as would be required using nucleic acid probes, since recombinant cDNA sequences may vary in orientation and fusion protein reading frame. Moreover, some fusion proteins may be lethal to the host cell, although in some cases this possibility may be controlled by propagating the cells prior to induction of beta-galactosidase fusion proteins on host cells that carry a repressor of lac Z gene transcription (Young and Davis, 1983b).

2.7. Analysis of Specific Clones

Regardless of the method used for selecting positive clones, several rounds of screening are normally required to identify and purify clones that reproducibly react with a given probe. Once reactivity with the probe has been established, the identity of a clone must be verified by more rigorous criteria, such as restriction enzyme mapping, hybrid-selected translation, subcloning into a vector containing a promoter for RNA polymerase and translating the transcription product of the insert, or nucleic acid sequence analysis.

The first step in the characterization of a selected recombinant normally involves determination of insert size and mapping of restriction enzyme recognition sites within the insert (Smith and Birnstiel, 1976). This information is useful for comparison with restriction maps of known DNA sequences and for subcloning phage:cDNA recombinants into a plasmid vector to enrich the cDNA:vector mass ratio and thus facilitate handling and analysis of the cloned insert. During electrophoretic analysis of the insert cDNAs, it is often convenient to transfer the DNA fragments from the gel to a solid membrane (Southern, 1975) for hybridization with a nucleic acid probe (if available) as a further confirmation of the identity and purity of the selected cDNA. Normally, the longest positively reacting clone is chosen for further analysis since the longest clones are the most likely to represent full-length mRNAs. However, the 5' end of the template mRNA may be represented by shorter clones because of incomplete second-strand synthesis or anomalous priming of the reverse transcriptase reaction. Thus, in the absence of a full-length clone, several shorter, overlapping clones may be used to derive information about a longer message. If the size of the naturally occurring mRNA has not been determined prior to cDNA library preparation, this information can be obtained by using the selected insert as a probe for Northern blot analysis. If the Northern blot reveals that the selected insert is much shorter than the full-length mRNA, the insert may be used as a cDNA probe to rescreen the library for recombinants containing longer inserts or to select overlapping clones. If rescreening is necessary, subcloning prior to insert/probe isolation is recommended, since insert preparations purified from a vector:cDNA recombinant molecule quite often crossreact with the parent vector, despite heroic efforts to obtain a population of insert molecules free of contaminating DNA sequences.

A useful technique in the characterization of selected clones is that of hybrid-selected translation. In this procedure, cDNA sequences are used to select specific mRNAs from a heterogeneous population by hybridization. The selected messages are then translated in vitro and the protein products are analyzed directly or by immunoprecipitation with a specific antibody (Parnes et al., 1981; Paterson and Roberts, 1981). When using this procedure, it is important to realize that a clone originally selected with an antibody probe must encode amino acid sequences recognized by that antibody, yet still may not represent the mRNA of interest. The demonstration that the cloned cDNA hybrid-selects mRNAs that direct the synthesis of proteins that crossreact with that same antibody is not sufficient proof of the identity of the clone. Characterization of the size or other physical characteristics of the proteins translated in vitro is thus recommended in addition to analysis of their immunoreactivity.

Although a number of other confirmatory tests can be run (e.g., see Kimmel, 1987), the final proof of identity of a given cDNA molecule must lie in an analysis of its nucleotide sequence. The nucleotide sequencing methods most often used today rely on subcloning into the M13 phage or the pUC plasmids, which contain a multipurpose cloning site adjacent to a sequencing primer recognition site (Heidecker et al., 1980; Messing et al., 1981; Vieira and Messing, 1982; Messing, 1983). DNA fragments are subcloned into the double-stranded replicative form of M13 and the resultant recombinants are then used to transfect competent host cells for propagation of single-stranded phage. The single-stranded recombinants then serve as templates for sequencing by the chain-termination procedure (Sanger et al., 1977). The terminal sequences of cDNAs cloned into the pUC plasmids can also be determined using the M13 universal sequencing primers. In this case, the recombinant plasmid DNA is converted to a single-stranded form by denaturation prior to annealing the primer. Directional subcloning into a sequencing vector is possible through the use of "paired" vectors so that the sequencing reaction can be deliberately primed from either end of a clone (Messing and Vieira, 1982). Using the M13 system, it is usually possible to derive 300–400 nucleotides of sequence information from a given reaction. In addition, several methods exist for extending sequence data to encompass several thousand nucleotides. The two most useful approaches involve the synthesis of oligonucleotides complementary to the 3' end of known, previously sequenced areas of the clone to direct the sequencing reaction through areas of particular interest (Barnes, 1987), and the preparation of nested, overlapping deletions of the parent clone to render various regions of the target sequence directly adjacent to the standard sequencing primer (Dale et al., 1985). The first of these two approaches can be quite expensive and/or labor intensive, depending on the availability of custom-synthesized oligonucleotides and the number of unique primers required for a given sequencing project. The second method allows for much more rapid data acquisition, but it introduces an element of randomness, since it is generally possible to engineer deletions only over a given range of nucleotides, rather than selectively to a specific region or nucleotide. Whenever possible, the sequence of a clone must be determined in both directions to clarify ambiguous sequence data introduced by structural anomalies in the template.

3. Methods

The cDNA synthesis protocol presented below is based on the technique described by Gubler and Hoffman (1982) and has been used successfully for the preparation of lambda gt11 libraries generated from mouse brain RNA (Newman et al., 1987) and fetal human spinal cord RNA (Roth et al., 1986). Although sufficient detail has been included so that each step in the preparation of a library may be replicated if so desired, this section is intended as a procedural guide rather than as a detailed laboratory manual. In addition to the library preparation protocol, general techniques for screening the library with antibody probes and characterization of selected clones by nucleic acid sequence analysis are included here. Other pertinent methods, such as screening by nucleic acid hybridization and purification of insert DNAs from recombinants, are presented elsewhere in this volume. For a presentation of other cloning and screening routes and a more extensive discussion of each step in the overall process, the reader is referred to Kimmel and Berger (1987).

3.1. Preparation of a cDNA Library in Lambda gt11

3.1.1. First Strand Synthesis

Poly (A)+ RNA is used as a template for synthesis of a cDNA molecule under the direction of reverse transcriptase.

3.1.1.1. Dilute the mRNA in H_2O to a concentration of 0.25 mg/mL or less, denature the mRNA by incubation at 70°C for 2–5 min, then transfer to ice.

3.1.1.2. On ice, prepare the following mixture in a 1.5 mL polypropylene test tube:

component	final concentration
Dithiothreitol (DTT)	1 mM
RNasin	400 U/mL
Tris-Cl, pH 8.3 at 42°C	100 mM
KCl	50 mM (final K^+ concentration)
$MgCl_2$	10 mM
Oligo(dT)$_{12-18}$	100 µg/mL
dNTP mixture	1 mM each dNTP
[alpha-^{32}P]dCTP, 3200Ci/mmol	400 µCi/mL
AMV reverse transcriptase	50 U/5 µg RNA
poly (A)+ RNA, 5 ug	100 µg/mL

Add H_2O to bring the final volume to 50µL.

Mix by tapping gently; centrifuge briefly to consolidate the reaction mixture in the base of the tube and to remove bubbles.

Incubate at 42°C for 1–3 h.

Stop the reaction by adding EDTA to 25µM.

NOTES: All solutions and labware must be treated to eliminate ribonuclease contamination. In subsequent steps, all labware coming into contact with the cDNA preparation should be siliconized prior to use (Maniatis et al., 1982) since single-stranded DNAs are notoriously "sticky" and may adhere quite tenaciously to the walls of test tubes and pipet tips.

The DTT must be added prior to addition of the RNasin to stabilize the latter component.

The pH of the Tris-Cl should be adjusted at 42°C, since the reverse transcriptase is quite sensitive to fluctuations in pH and since the pH of Tris changes with temperature.

The contribution of the AMV reverse transcriptase storage buffer to the final monovalent cation concentration must be considered. The amount of KCl added to the reaction mixture will, therefore, vary according to the potassium ion concentration of the storage buffer. The AMV reverse transcriptase may be diluted prior to use to reduce the amount of glycerol added to the reaction mixture from the storage buffer.

Actinomycin D or pyrophosphate may be included in the reaction mixture to inhibit formation of hairpin loop structures in the newly synthesized cDNA (Krug and Berger, 1987).

dNTPs are prepared and stored in 10 mM Tris, pH 8.0 at a concentration of 20 mM or higher, as determined by optical density measurements at the wavelengths and extinction coefficients recommended by Maniatis et al. (1982). The stock dNTP mixture is diluted, if necessary, just prior to use.

[alpha-^{32}P]dCTP is included as a means of monitoring the progress of library preparation. Alternatively, the labeled nucleotide can be included in a second, smaller-scale reaction mixture prepared by transferring an aliquot of the primary mixture to a tube containing the [alpha-^{32}P]dCTP. The radioactively labeled reaction products are then used to follow the success of individual steps in the procedure and the cDNA synthesized for cloning purposes is left unlabeled.

3.1.1.3. Evaluate first strand synthesis by taking 1 μL aliquots of the reaction mixture at the beginning and end of the incubation and measuring total radioactivity in the reaction mixture and incorporation of radioactivity into TCA-precipitable material (Maniatis et al., 1982). Although analysis of the radioactive reaction products by agarose gel electrophoresis may be included at this point to evaluate the synthesis of full-length cDNAs (Krug and Berger, 1987), particularly when the technique is being executed for the first time, we routinely rely on the information provided by the TCA precipitation step and reserve electrophoretic analysis until after second strand synthesis (*see below*, Section 3.1.2.6.),

3.1.1.4. Extract the sample once with an equal volume of phenol:chloroform:isoamyl alcohol (24:24:1, v/v/v) and then re-extract the organic phase twice with one volume 10 mM Tris-C1, pH 8.0, and pool the aqueous phases (Maniatis et al., 1982). Remove residual phenol from the pooled aqueous phases by extracting twice with two volumes of ether, and allow the residual ether to evaporate under a hood. Add one volume of 4 M ammonium acetate and two volumes of chilled ethanol, and precipitate the cDNA:RNA hybrids at –70°C for 20 min or –20°C overnight. Pellet the cDNA:RNA hybrids by centrifugation, wash the pellet once with 80% ethanol, and dry under vacuum for 5–10 min (Maniatis et al., 1982). Prolonged drying of the pellet should be avoided.

NOTE: It is often helpful to obtain a rough idea of the radioactivity in the pellet at this point using a Geiger counter. The cDNA can then be followed through subsequent manipulations such as resuspension of dry pellets in aqueous solution and transfer of the sample from one tube to another. These manipulations can prove to be quite troublesome because of adherence of the cDNA to test tubes and pipet tips, even when using siliconized labware.

3.1.2. Second Strand Synthesis

This protocol, first described by Okayama and Berg (1982), and later modified by Gubler and Hoffman (1983), relies on the endonucleolytic properties of *E. coli* RNase H to introduce nicks in the RNA strand of the RNA:cDNA hybrid. The resulting RNA fragments, which are still attached to the cDNA, are used as primers for synthesis of a second DNA strand by DNA polymerase I.

3.1.2.1. Resuspend the first strand pellet in Tris-C1, pH 7.5.

NOTE: No more than 500 ng of single-stranded DNA (i.e., 1 µg cDNA:RNA hybrid) should be added to a 100 µL second-strand synthesis reaction. If more material is to be included, the reaction may be scaled up accordingly. We find it convenient to dissolve 1ug cDNA:RNA hybrid in 82 µL of 25 mM Tris-C1, pH 7.5, remove 2 µL for electrophoretic analysis, and use the remaining 80 µL for the second-strand reaction for a final Tris-C1 concentration of 20 mM in 100 µL.

3.1.2.2. After removing an aliquot for electrophoretic analysis, add the following reaction components to the cDNA:RNA hybrid.

component	final concentration
$MgCl_2$	5 mM
KCl	100 mM
$(NH_4)_2SO_4$	10 mM
Bovine Serum Albumin (BSA)	50 µg/mL
beta-NAD	0.15 mM
dNTP mixture	50 µM each dNTP
[alpha-^{32}P]dCTP,3200 Ci/mmol	200 µCi/mL
E. coli RNase H	8.5 U/mL
E. coli DNA ligase	10 U/mL
DNA polymerase I	230 U/mL

Add H_2O to bring the final volume to 100 µL.
Mix; centrifuge briefly.
Incubate at 12°C for 1 h and then at 22°C for 1 h.
Stop the reaction by adding EDTA to 25 mM.

3.1.2.3. Evaluate second strand synthesis by TCA precipitation of the radioactive reaction products (Section 3.1.1.3.). Again, a separate radioactive reaction may be carried out alongside a primary, unlabeled reaction if so desired. Generally, it is best to avoid incorporation of radioactive nucleotides into the first and second strands if prolonged storage of reaction products before proceeding with subsequent steps is anticipated. However, since the cDNA cloning process can be routinely completed in about 2 w, it is often most convenient to prepare a single radioactive reaction mixture for each step.

3.1.2.4. Extract the reaction mixture and precipitate the reaction products as in Section 3.1.1.4.

3.1.2.5. Resuspend the ds-cDNA pellet in 50 µL 10 mM Tris-Cl, pH 8.0, and remove 1.5 µL for electrophoretic analysis. To the remaining 48.5 µL add ammonium acetate and ethanol to precipitate DNA as indicated in Section 3.1.1.4.

3.1.2.6. Subject the reserved aliquots of the first and second strand synthesis reactions to electrophoresis through a 1.2% agarose gel (Maniatis et al., 1982). Include DNA size markers in adjacent lanes for determination of the cDNA size range. After electrophoresis is complete, stain the gel with ethidium bromide (0.5 µg/mL) to visualize the DNA markers, then fix the gel in 7.5% TCA for 30 min. Dry the fixed gel under two sheets of Whatman 3MM paper topped by a stack of paper towels and a weight (500 g–1 kg). Visualize the radioactive reaction products in the gel by autoradiography (Maniatis et al., 1982). The cDNAs should appear as a smear of radioactivity representing molecules of various different sizes.

3.1.3. EcoRI Site Methylation

This cloning protocol employs the use of synthetic EcoRI linkers to facilitate ligation of the cDNAs to lambda gt11 arms. Since the ds-cDNAs may have internal EcoRI recognition sites, they are treated with EcoRI methylase at this point to prevent cleavage with the enzyme during subsequent digestion of linker concatamers to generate cohesive ends.

3.1.3.1.   Resuspend the ds-cDNA in ~70 µL of H_2O.
3.1.3.2.   Add the folowing reaction components.

component	final concentration
Tris-Cl, pH 8.0	100 mM
EDTA, pH 8.0	10 mM
BSA	50 µg/mL
S-adenosyl methionine (stored in 0.01 N H_2SO_4)	50 µM
EcoRI methylase	60 U/mL

Add H_2O to bring the final volume to 100 µL.
Mix; centrifuge briefly.
Incubate at 37°C for 1 h.
Stop the reaction by incubation at 70°C for 5 min.

3.1.3.3.   Extract the reaction mixture and precipitate the reaction products as in Section *3.1.1.4*.

NOTE: The success of the methylation reaction can be tested by setting up a parallel reaction mixture using 0.5–2.0 µg of a lambda DNA standard in place of the synthetic ds-cDNA. The methylated lambda DNA is then assayed by exposure to EcoRI in a standard EcoRI digestion reaction (Maniatis et al., 1982). Uncut lambda DNA, EcoRI-digested lambda DNA, and methylated EcoRI-treated lambda DNA are then subjected to agarose gel electrophoresis. The methylated EcoRI-treated lambda DNA will migrate alongside the uncut lambda DNA if methylation of EcoRI sites was complete.

3.1.4. Fill-in Reaction

Before the addition of synthetic linkers, the ds-cDNA must be made blunt-ended. Although this can be accomplished using S1 nuclease or T4 DNA polymerase, we have achieved greater success using the Klenow fragment of DNA polymerase I.

3.1.4.1. Resuspend the ds-cDNA in 20 µL of 1 mM Tris-Cl, pH 7.5; 0.1 mM EDTA.

3.1.4.2. Add the following reaction components.

component	final concentration
Tris-Cl, pH 7.2	40 mM
MgSO$_4$	8 mM
DTT	80 µM
dNTP mixture	80 µM each dNTP
BSA	50 µg/mL
Klenow fragment of DNA polymerase I	40 U/mL

Add H$_2$O to bring the final volume to 50 µL.
Mix; centrifuge briefly.
Incubate at 22°C for 30 min.
Stop the reaction by adding EDTA to 20 mM.

3.1.4.3. Extract the reaction mixture and precipitate the reaction products as in Section *3.1.1.4*.

3.1.5. Linker-cDNA Ligation

Synthetic oligonucleotide linkers are added to the blunt-ended ds-cDNA with T4 DNA ligase, and subsequently digested with EcoRI to generate cohesive ends. It is advisable to prepare and test the linkers ahead of time so that they will be ready for use as soon as the blunt-ended ds-cDNA has been prepared. Dephosphorylated EcoRI linkers (e.g., the 10 bp oligomers supplied by New England Biolabs) are first phosphorylated ("kinased") using T4 polynucleotide kinase and [gamma-^{32}P]ATP so that subsequent steps can be monitored.

3.1.5.1. Dissolve 1 OD$_{260}$ unit (~50ug DNA) dephosphorylated EcoRI linkers in 100 µL of 10 mM Tris-Cl, pH 7.5.

3.1.5.2. Combine the following reaction components.

component	final concentration
Linkers	0.2 mg/mL
Tris-Cl, pH 7.5	66 mM

cDNA Library Preparation

component	final concentration
MgCl$_2$	10 mM
DTT	5 mM
[gamma-^{32}P]ATP (1000–3000 Ci/Mmol)	2 mCi/mL
T4 DNA kinase	1000 U/mL

Add H$_2$O to bring the final reaction volume to 20 µL.
Incubate at 37°C for 15 min.

3.1.5.3. Add the following components to the reaction mixture.

component	final concentration
Tris-Cl, pH 7.5	66 mM
MgCl$_2$	10 mM
DTT	5 mM
ATP	1 mM
T4 DNA kinase	1000 U/mL

Add H$_2$O to bring the final, combined reaction volume to 40 µL.
Mix; centrifuge briefly.
Incubate at 37°C for 30 min.
Store the kinased linkers at –20°C until ready for use. The final linker concentration is 0.1 µg/µL in 40 µL.

3.1.5.4. To test the ligation of the linkers, prepare the following reaction mixture.

component	final concentration
Kinased linkers	0.01 µg/uL
Tris-Cl, pH 7.5	66 mM
MgCl$_2$	5 mM
DTT	5 mM
ATP	1 mM
T4 DNA ligase	600 Weiss U/mL

Add H$_2$O to bring the final reaction volume to 10 µL.
Mix; centrifuge briefly.
Incubate at 12–14°C for 4–16 h.
Stop the reaction by incubation at 65°C for 10 min.
NOTE: The units used to measure T4 DNA ligase may vary according to manufacturer (e.g., 1 Weiss U is equivalent to 0.015 U of activity, as defined by New England Biolabs).

3.1.5.5. Transfer 5 μL of the ligated kinased linkers to a separate tube and digest with 10–15 U of EcoRI in a 20 μL reaction volume.

3.1.5.6. Analyze 0.5 μL unligated kinased linkers, 5 μL ligated kinased linkers (representing 0.5 μL kinased linkers), and 20 μL EcoRI-digested ligated kinased linkers (representing 0.5 μL kinased linkers) by electrophoresis through an 8% polyacrylamide gel (Maniatis et al., 1982). On autoradiographic analysis, the ligated kinased linkers should appear as a ladder of concatameric molecules, whereas the unligated kinased linkers and the EcoRI-digested ligated kinased linkers should both appear as a band of monomers at the base of the gel.

3.1.5.7. Ligate the kinased linkers to the ds-cDNA by adding the following components to the dry ds-cDNA pellet.

component	final concentration
Kinased linkers	50 μg/mL
Tris-Cl, pH 7.5	33 mM
$MgCl_2$	5 mM
DTT	10 mM
ATP	1 mM
T4 DNA ligase	600 Weiss U/mL

Add H_2O to bring the final volume to 10 μL.

Mix; centrifuge briefly; remove 0.25 μL and reserve for electrophoretic analysis.

Incubate at 12–14°C for 4–16 h.

Stop the reaction by incubation at 65°C for 10 min.

3.1.5.8. Digest the ligated ds-cDNA:linkers with 25–30 U of EcoRI in a final volume of 20 μL, reserving 0.5 μL of the initial EcoRI reaction mixture for electrophoretic analysis. Terminate the digestion reaction by incubation at 70°C for 10 min, and remove an additional 0.5 μL for electrophoretic analysis.

3.1.5.9. Precipitate the reaction products as in Section *3.1.1.4*.

3.1.5.10. Repeat the ethanol precipitation once.

3.1.5.11. Analyze the reserved unligated, ligated, and ligated EcoRI-digested ds-cDNA:linker samples by electrophoresis through an 8% polyacrylamide gel and visualize the reaction products on the gel by autoradiography as in step *3.1.5.6*. Because the linker fragments contribute much more radioactivity to the sample

cDNA Library Preparation

than the cDNA, the bulk of the radioactivity in the unligated and ligated EcoRI-digested samples will migrate as a discrete band near the base of the gel, whereas the ligated cDNA:linker molecules will appear as a diffuse smear originating at the top of the gel.

3.1.6. Purification of cDNAs by Column Chromatography

This step accomplishes two goals: removal of excess linker fragments, and selection of ds-cDNAs according to size. Detailed instructions for column preparation and calibration have been published (Huynh et al., 1985; Eschenfeldt and Berger, 1987). The procedure we have used is as follows.

3.1.6.1. Siliconize a 1ml glass pipet and plug the tip with siliconized glass wool. Wash the pipet sequentially with H_2O and ethanol, then autoclave and dry.

3.1.6.2. Wash hydrated Sepharose CL-4B several times in column buffer (10 mM Tris-C1, pH7.5; 100 mM NaCl; 1 mM EDTA), degas, and pour the slurry into a 10 mL syringe attached to the column filled with column buffer.

3.1.6.3. Wash the packed column with at least 50 mL column buffer to remove contaminants that inhibit ligation (this is usually accomplished by washing overnight).

3.1.6.4. Precalibrate the column, if desired, by running radioactive DNAs of known length (e.g., restriction fragments) through the column and analyzing the collected fractions by electrophoresis and autoradiography. We often skip this step and instead perform a similar analysis using a portion of each cDNA fraction eluted from the column.

3.1.6.5. Dissolve the ligated linker:cDNAs in 30 μL TE (10 mM Tris-C1, pH 7.5; 1 mM EDTA) and apply the sample to the column. Allow the sample to flow into the column, collecting the eluate into 1.5 mL polypropylene test tubes. Rinse the sample tube with 20 μL TE and apply the rinse to the column. After the rinse has run into the column, fill the column and the attached reservoir with column buffer and commence collecting 1 drop fractions (~45 μL each). Monitor the radioactivity in the collected samples using a Geiger counter or by measuring Cerenkov radiation.

3.1.6.6. Analyze aliquots corresponding to ~500cpm from selected fractions by electrophoresis through a 1.5% agarose gel (including a DNA size standard on the gel) followed by autoradiography. The fractions to be analyzed are selected based on the radioactivity in each sample and are usually confined to those fractions that are excluded from the column. The leading edge of the radioactive peak will contain the longer molecules.

3.1.6.7. Combine the fractions containing ds-cDNAs that fall into the desired size range.

3.1.7. Lambda gt11-cDNA Ligation

The EcoRI-digested linker-cDNA molecules are joined to lambda gt11 arms containing EcoRI-generated cohesive termini that have been dephosphorylated to prevent self-ligation.

3.1.7.1. Add 1 µg EcoRI-digested dephosphorylated lambda gt11 (e.g., as supplied in the Protoclone lambda gt11 system available from Promega Biotec) to the ds-cDNAs selected by column chromatography and precipitate the mixture, as in Section 3.1.1.4.

3.1.7.2. To the vector-cDNA pellet, add 7.6 µL 13.15 mM Tris-CL, pH7.5; 13.15 mM MgCl$_2$ and dissolve.

3.1.7.3. Incubate at 42°C for 15 min to ensure that complementary ends are annealed.

3.1.7.4. Add the following reaction components:

component	final concentration
ATP	1 mM
DTT	10 mM
T4 DNA ligase	600 Weiss U/mL

Add H$_2$O to bring the final volume to 10 µL.
Mix; centrifuge briefly.
Incubate at 12–14°C 4–16 h.

3.1.8. Packaging of Recombinant Lambda gt11:cDNAs

The recombinant molecules are packaged into infectious particles in vitro. We have successfully used the prepared packaging mix available from Promega Biotec.

3.1.8.1. Package the recombinants according to the instructions provided by the supplier of the packaging extract. This usually involves mixing the ligated cDNA:vector with the packaging extract as it thaws and incubating the mixture at 22°C for 2–3 h.

3.1.8.2. Add phage dilution buffer (100 mM NaCl; 10 mM Tris-C1, pH 7.9; 10 mM MgSO$_4$) to a final volume of 0.5 mL and add 25 µL of chloroform. This mixture constitutes the cDNA:lambda gt11 library. The library may be stored at 4°C prior to titer determination, amplification, and screening.

cDNA Library Preparation

3.1.9. Determination of Library Titer

The number of recombinant phage in the library is determined by plating on the bacterial host strain Y1090(hsdR-). This is a restriction minus derivative of the original Y1090 host strain used by Young and Davis (1983b), a characteristic that obviates the need to plate and amplify the library on Y1088 before screening. Another notable feature of Y1090 genetics is that this host contains a deletion of the lac operon, permitting the detection of recombinants based on the absence of beta-galactosidase activity. Using Y1090 as the host strain, the recombinant phage will appear as clear plaques on a background of blue (nonrecombinant) plaques when plated in the presence of 5-bromo-4-chloro-3-indolyl beta-D-galactoside (X-gal), the chromogenic substrate for beta-galactosidase. Isopropyl-beta-D-thiogalactopyranoside (IPTG) is included in the culture medium to induce lac Z gene transcription.

3.1.9.1.   Combine 100 µL dilutions containing 0.5 and 5.0 µL of the library with 100 µL Y1090 plating cells (Maniatis et al., 1982) and let the mixture stand at room temperature for 15–20 min.

3.1.9.2.   Plate the library on dry 90 mm LB plates in 2.5 mL LB top agar (Maniatis et al., 1982) containing 20 µL of 100 mM IPTG and 40 µL of 20 mg/mL X-gal (prepared in dimethylformamide).

3.1.9.3.   Incubate overnight at 37°C.

3.1.9.4.   Score plates for the number of clear plaques relative to the total number of plaques to obtain the percentage of recombinants in the library. Extrapolate the library titer based on the number of clear plaques per volume of library plated.

3.2. Amplification and Screening of the Completed Library

3.2.1. Amplification

The recombinant phage are propagated prior to screening and characterization. Y1090(hsdR-) is again used as the host strain. To prevent introduction of a bias against recombinants that produce fusion proteins detrimental to host cell survival, the Y1090 bacteria carry a repressor of lac Z gene transcription.

3.2.1.1.   Prepare plating mixtures containing 10,000–50,000 recombinants and 100 µL plating bacteria as above and plate on dry 150 mm LB plates in 6 mL LB top agar (IPTG and X-gal are not included here). Plate the entire library.

3.2.1.2.   Incubate overnight at 37°C.

3.2.1.3. Add 11 mL cold SM (100 mM NaCl; 50 mM Tris-Cl, pH 7.5; 8 mM MgSO$_4$; 0.01% gelatin) and incubate with gentle agitation overnight at 4°C.

3.2.1.4. Remove the SM gently to a sterile 50 mL screw-capped centrifuge tube and wash the plate with an additional 4 mL of SM.

3.2.1.5. Pool the SM plate washes and add chloroform to 5%.

3.2.1.6. Pellet cellular debris by centrifugation at 4000x g for 5 min at 4°C.

3.2.1.7. Transfer the supernatant to a clean tube and add chloroform to 0.3%. The library may now be stored indefinitely at 4°C.

3.2.2. Screening the Library for Specific Clones

The library may be screened in Y1090 using either nucleic acid or antibody probes. Since detailed protocols for screening with antibody probes have been published (Young and Davis, 1983b; Huynh et al., 1985; Mierendorf et al., 1987), only the salient features of the process will be presented here.

3.2.2.1. Again, Y1090 is the host bacteria of choice. The cultures are incubated just until plaques appear on a smooth bacterial lawn, and then IPTG-soaked nitrocellulose filters are gently laid on the surface of the plates. In this way, the lac repressor carried by the Y1090 host suppresses expression of potentially detrimental fusion proteins during bacterial growth and plaque formation. The lac Z gene is then derepressed by IPTG, permitting synthesis of foreign proteins. Proteolytic degradation of expressed proteins is reduced in Y1090 cells due to a deficiency in the lon protease.

3.2.2.2. After 3–8 h, the filters are removed from the plates, washed, treated to block nonspecific protein binding sites, and incubated in a solution of primary antibody. The primary antibody must be pretreated to remove IgGs that recognize *E. coli* proteins (Huynh et al., 1985).

3.2.2.3. The primary antibody-antigen complex may be detected using ^{125}I-labeled protein A (Young and Davis, 1983b; Huynh et al., 1985) or a second antibody conjugated with alkaline phosphatase or peroxidase (Mierendorf et al., 1987).

3.2.2.4. Plaques corresponding to positive signals on the nitrocellulose replicas are picked from the master plate and replated and screened again. This process is repeated until all of the plaques on a given plate produce a positive signal, indicating that a pure culture of the recombinant organism has been obtained.

3.3. Characterization of Positive Clones

Once the bacteriophage clone is isolated, the cDNA insert can be analyzed by restriction mapping techniques and/or subcloned into an appropriate plasmid vector for future analysis.

The nucleotide sequence of the clone is derived primarily from methods employing chemically induced cleavage reactions (Maxam and Gilbert, 1977) or chain termination reactions that use dideoxyribonucleotides to block elongation of DNA molecules as they are synthesized (Sanger et al., 1977). Specialized vectors, such as M13, greatly facilitate the sequencing process, although their use does normally require an increase in the number of manipulations that must be performed on a cDNA before obtaining sequence information. Once a cDNA has been subcloned into M13, template DNA for sequencing can usually be purified in a day and the sequencing reactions run on the following day. As mentioned above, 300-400 bp of nucleotide sequence data can routinely be obtained from a single sequencing reaction. The preparation of deletion mutants of the parent M13 recombinant to generate overlapping clones permits sequence analysis of clones several thousand nucleotides in length.

Dideoxyribonucleotide sequence analysis can be performed using any one of a number of protocols, each with its own idiosyncracies. Whatever approach one chooses, the final measure of success lies with the quality of the sequence data obtained. Preparation of template DNA of high purity and judicious attention to detail in preparing and handling the sequencing gels can eliminate many of the most common sources of error.

A number of commercially prepared kits are available to facilitate sequencing in M13, and extensive instructions are usually provided with each system. We have successfully used the M13 Cloning/Dideoxy Sequencing kit provided by Bethesda Research Laboratories (BRL) with minor modifications, as listed below. Deletion mutants of the M13 clones are routinely prepared using the Cyclone system of International Biotechnologies, Inc. (IBI).

A comprehensive discussion of the various approaches to DNA sequence analysis, including theoretical considerations and troubleshooting advice, is presented by Hindley (1983), and the M13 cloning and sequencing vectors are reviewed by Messing (1983). For more detailed protocols for dideoxy sequencing using [alpha-^{35}S]dATP, the reader is referred to Williams et al. (1986) and to Barnes (1987).

3.3.1. Preparation of Template DNA for Dideoxy Sequencing

3.3.1.1. Subclone the cDNA insert into a single restriction site in the multipurpose cloning region of M13mp19 or subclone by forced orientation using the paired vectors M13mp18 and M13mp19. The choice of cloning site depends on the insert to be cloned. We routinely subclone directly from the EcoRI site of lambda gt11 into the EcoRI site of M13mp19. In this case, both orientations are possible and several subclones must be chosen to obtain sequence data from both ends of the clone.

3.3.1.2. Select clones that contain inserts as indicated by the loss of alpha-complementation of the host cell's beta-galactosidase gene (i.e., clones that form clear plaques when plated in the presence of X-gal and IPTG). Plaques may be screened by nucleic acid hybridization, if so desired (Grunstein and Hogness, 1975; Hanahan and Meselson, 1980). Selection of specific clones by hybridization is especially useful when subcloning a mixture of fragments derived from restriction enzyme digestion of a recombinant cDNA, since in this case, the M13 library will consist of a heterogeneous population of recombinant cDNA:M13 molecules. When subcloning directly from lambda gt11, it is not necessary to purify the inserts from the EcoRI-cleaved lambda gt11, since the 50 kb vector molecule will not be accepted as an insert in M13.

3.3.1.3. Transfer selected plaques into 1.5 mL of 2xYT broth and add 15 µL of an exponentially growing culture of JM103 (2xYT consists of 16 g Bactotryptone, 10 gm Bacto-yeast, and 10 gm NaCl in 1 l).

3.3.1.4. Incubate in 18 × 150 mm sterile glass tubes for 5 h at 37°C with shaking at 250 rpm.

3.3.1.5. Transfer the cultures to a 1.5 mL polypropylene test tube and precipitate the cellular debris by centrifugation at ~12,000xg for 15–20 min.

3.3.1.6. Transfer 1 mL of the supernatant to a fresh tube, avoiding the cell pellet. Add 250 µL of 20% polyethylene glycol-2.5M NaCl and incubate at room temperature for 15 min.

3.3.1.7. Pellet the phage particles by centrifugation at ~12,000xg for 5 min. Remove the supernatant, recentrifuge briefly, and remove the residual supernatant from the second spin (it is very important at this point to have a clean pellet).

3.3.1.8. Add 100 µL TES (20 mM Tris-Cl, pH7.5; 0.1 mM EDTA; 10 mM NaCl) to the pellet and extract with 50 µL phenol. After removing the organic layer, reextract the aqueous phase with ether to remove residual phenol.

3.3.1.9. Add 8 µL 3M sodium acetate, pH 5.2, and 200 µL ethanol and precipitate at –70°C for 15–20 min, followed by centrifugation at ~12,000xg for 15 min. Discard the supernatant.

3.3.1.10. Resuspend the dry pellet in 25 µL TES. The preparations may be stored at –20°C until needed. It is advisable to analyze about 1.5 µL of each preparation by electrophoresis through a 0.7% agarose gel to check the yield of single-stranded DNA. By also running a single-stranded preparation from a blue (nonrecombinant) plaque on the gel, the presence of inserts in the M13 can be ascertained by direct comparison of size.

3.3.1.11. When subcloning into a single restriction site (such as EcoRI) in M13, the DNA may be inserted in either direction, yielding single-stranded preps with sense or anti-sense cDNA inserts. To select clones for sequencing that have been inserted in opposite directions, a "C-test" may be performed at this point (Messing, 1983). The C-test is based on the fact that the M13mp19 sequence in all the single-stranded preparations is identical and noncomplementary, and, therefore, only those clones possessing complementary inserts will hybridize to one another. By selecting single-stranded M13 clones with complementary inserts for sequence analysis, information can be derived from both ends of the cDNA insert using the same sequencing primer.

3.3.2. Prepare Solutions for the Dideoxy Sequencing Reaction

3.3.2.1. Prepare polymerase reaction buffer (PRB) (100 mM Tris-Cl, pH8.0; 50 mM MgCl$_2$) for reactions using ^{35}S dATP as the label. Store in 1 mL aliquots at –20°C.

3.3.2.2. Deoxynucleotides (dNTPs) are maintained as 10 mM stocks, and may be stored as 0.5 mM working stocks at –20°C for 2 wk–1 mon. The following mixtures should be freshly prepared every 2 wk.

$A°$ mix
10 µL 0.5 mM dTTP
10 µL 0.5 mM dCTP
10 µL 0.5 mM dGTP
10 µL PRB

$C°$ mix
10 µL 0.5 mM dTTP
0.5 µL 0.5 mM dCTP
10 µL 0.5 mM dGTP
10 µL PRB

G^o mix
10 μL 0.5 mM dTTP
10 μL 0.5 mM dCTP
0.5 μL 0.5 mM dGTP
10 μL PRB

T^o mix
0.5 μL 0.5 mM dTTP
10 μL 0.5 mM dCTP
10 μL 0.5 mM dGTP
10 μL PRB

Store at –20°C.

3.3.2.3. Dideoxynucleotide stocks (ddNTPs) are stored at a concentration of 10 mM at –20°C. The working stocks (below) must be freshly prepared every 2 wk.

ddA = 0.1 mM
ddC = 0.1 mM
ddG = 0.1 mM
ddT = 0.4 mM

The optimal concentration of the ddNTP working stocks may vary, and must be determined empirically for each sequencing project.

3.3.2.4. Formamide Dye Mix

0.1% Xylene cyanole (w/v)
0.1% Bromophenol blue (w/v)
10 mM Na$_2$ EDTA
95% deionized formamide (v/v)

3.3.3. Dideoxy Sequencing Reactions Using [alpha-^{35}S]dATP

3.3.3.1. Anneal the sequencing primer to the template DNA. In a 1.5 mL polypropylene test tube, combine the following components.

5.0 μL template DNA (0.5–1.0 μg)
2.0 μL 17 bp M13 universal sequencing primer (4 ng, supplied by BRL)
1.0 μL PRB
4.5 μL H$_2$O

Place the mixture in a heating block filled with H$_2$O preheated to 90°C. Incubate the mixture for 5 min, then remove the block from the heating apparatus, and allow the block and the tubes it contains to cool slowly to room temperature (this usually takes about 1.5 h).

3.3.3.2. While the annealing mixture is cooling, prepare the following reaction tubes.

cDNA Library Preparation

A) 1 μL A° mix + 1 μL ddA
C) 1 μL C° mix + 1 μL ddC
G) 1 μL G° mix + 1 μL ddG
T) 1 μL T° mix + 1 μL ddT

3.3.3.3. After the annealing reaction is complete, centrifuge the tubes briefly to consolidate the reaction components in the bottom of the tube and add the following components directly to the annealed mixtures.

3 μL [alpha-^{35}S]dATP (10 μCi/μL, 600Ci/mmol)
1.0 μL 0.1 M DTT
1.0 μL Klenow fragment of DNA polymerase I (3 U/μL)

Dispense 3 μL aliquots of the above mixture to each of the four reaction tubes (A,C,G,T) and mix by gently pipeting the mixtures up and down.
Incubate at 30°C for 20 min.

3.3.3.4. Add 1.0 μL 0.5 mM ATP to all tubes. Mix gently and incubate at 30°C for 20–25 min.

3.3.3.5. Stop the reaction by adding 5 μL formamide dye mix to each tube. The reaction mixtures may be stored for up to a week at –20°C before electrophoretic analysis.

3.3.4. Electrophoretic Analysis of the Sequencing Reaction Products

To obtain the optimal amount of readable sequence, it is necessary to run the samples through two separate (8% and 6%) gel systems. We usually run the 8% gel for ~3.5 h at constant power (60W) and can reliably read 180–250 bp from the resulting autoradiograph. The 6% gel is run under the same conditions for ~12 h to overlap with the 8% gel around bp# 150–180, and increase the amount of sequence information obtained from a single reaction to as high as 500bp.

A recipe for preparing the sequencing gels is presented below. A 60 cm sequencing apparatus is used with tapered spacers and a sharkstooth comb. One of the gel plates is siliconized prior to pouring the gel, and both gel plates must be absolutely clean. We prepare the sequencing apparatus for electrophoresis and fix the gel for autoradiography as described in Williams et al. (1986).

3.3.4.1. Prepare the following gel mixtures.

component	6% Gel	8% Gel
Urea	105 g	84 g
Acrylamide	14.25 g	15.2 g
Bis-acrylamide	0.75 g	0.8 g
BioRex resin	12.5 g	10.0 g
10 x TBE	25 mL	20 mL
Final volume	250 mL	200 mL

5 mL above mix + 50 µL 10% ammonium persulfate (APS) + 6 µL TEMED

10% APS	2.38 mL	1.9 mL
TEMED	100 µL	80 µL

Combine Urea, Acrylamide, and Bis-acrylamide and dissolve in H_2O in a warm H_2O bath.

Add Bio-Rex resin (BioRad) and deionize, stirring gently, for 20 min.

Filter the deionized mixture through Whatman #1 filter paper.

Add the 10x TBE *(see formula below)* adjust the final volume, and degas the mixture for 15–20 min.

Transfer 5 mL of the gel mixture to a small beaker and add ammonium persulfate and TEMED as above. Use this 5 mL mixture to pour a "plug," sealing the base of the gel. Let the plug polymerize ~5 min.

Add the indicated amounts of APS and TEMED to the remaining mixture and pour the gel.

10x TBE is 0.5M Tris base; 0.5M Boric acid; 10 mM EDTA; pH 8.3.

3.3.4.2. Place the comb in the gel apparatus flat side down to form a flat surface on the top of the gel.

3.3.4.3. After the gel has polymerized (~ 1h), briefly remove the clamps that hold the apparatus together to place one layer of 1x TBE-soaked Whatman 3MM paper and one layer of plastic wrap over the edges of the gel. This precaution is necessary only if the gel is to be stored overnight before running, because keeping the gel moist seems to prevent channeling along the sides and bottom. Gels that are run on the same day as they are poured run a bit faster than those allowed to polymerize overnight, as judged by the migration of the tracking dyes.

3.3.4.4. Before running the gel, reverse the comb to form sample lanes and wash the surface of the gel thoroughly with Ix TBE. The samples are boiled for 3 min and quenched on ice just prior to loading them on the gel. The running buffer is Ix TBE.

The gel can be read by eye or by using a gel reader interfaced with a computer (e.g., the IBI-Pustell system). For hints on reading the gels and advice on reading through ambiguous sequences, *see* Hindley (1983).

Acknowledgments

The authors wish to thank Thea Anzalone for assistance in the preparation of the manuscript. This work was supported, in part, by National of Institutes of Health Grants NS 23022, NS 23322, and NS25304.

References

Bahl C. P., Marians K. J., Wu R., Stawinsky J., and Narang S. A. (1976) A general method for inserting specific DNA sequences into cloning vehicles. *Gene* **1**, 81–92.

Baltimore D. (1970) RNA-dependent DNA Polymerase in virions of RNA tumour viruses. *Nature* **226**, 1209–1211.

Barnes W. M. (1987) Sequencing DNA with dideoxyribonucleotides as chain terminators: Hints and strategies for big projects, in *Methods in Enzymology, vol. 152: Guide to Molecular Cloning Techniques* (Berger S. L. and Kimmel A. R., eds.) pp. 538–556, Academic Press, New York.

Benton W. D. and Davis R. W. (1977) Screening lambda gt recombinant clones by hybridization to single plaques *in situ. Science* **196**, 180–182.

Berger S. L. and Kimmel A. R. (eds.) (1987) *Methods in Enzymology, vol. 152: Guide to Molecular Cloning Techniques.* Academic Press, New York.

Bolivar F., Rodriguez R. L., Greene P. J., Betlach M. C., Heyneker H. L., Boyer H. W., Crosa J. H., and Falkow S. (1977) Construction and characterization of new cloning vehicles. II. A multipurpose cloning system. *Gene* **2**, 95–113.

Chang A. C. Y. and Cohen S. N. (1974) Genome construction between bacterial species *in vitro:* replication and expression of *Staphylococcus* plasmid genes in *Escherichia coli*. *Proc. Natl. Acad. Sci. USA* **71**, 1030–1034.

Cohen S. N. and Chang A. C. Y. (1973) Recircularization and autonomous replication of a sheared R-factor DNA segment in *Escherichia coli* transformants. *Proc. Natl. Acad. Sci. USA* **70**, 1293–1297.

Cohen S. N., Chang A. C. Y., and Hsu L. (1972) Nonchromosomal antibiotic resistance in bacteria: Genetic transformation of *Escherichia coli* by R-factor DNA. *Proc. Natl. Acad. Sci USA* **69**, 2110–2114.

Cohen S. N., Chang A. C. Y., Boyer H. W., and Helling R. B. (1973) Construction of biologically functional bacterial plasmids *in vitro*. *Proc. Natl. Acad. Sci. USA* **70**, 3240–3244.

Collins J. and Hohn B. (1978) Cosmids: A type of plasmid gene-cloning vector that is packageable *in vitro* in bacteriophage lambda heads. *Proc. Natl. Acad. Sci. USA* **75**, 4242–4246.

Dale R. M. K., McClure B. A., and Houchins J. (1985) A rapid single-stranded cloning strategy for producing a sequential series of overlapping clones to use in DNA sequencing. Applications to sequencing the corn mitochondrial 18S rDNA. *Plasmid* **13**, 31–40.

Efstratiadis A., Kafatos F. C., Maxam A. M. and Maniatis T. (1976) Enzymatic *in vitro* synthesis of globin genes. *Cell* **7**, 279–288.

Eschenfeldt W. H. and Berger S. L. (1987) Purification of large double-stranded cDNA fragments, in *Methods in Enzymology, vol. 152: Guide to Molecular Cloning Techniques* (Berger S. L. and Kimmel A. R., eds.), pp. 335–337, Academic Press, New York.

Gefter M. L., Becker A., and Hurwitz J. (1967) The enzymatic repair of DNA, I. Formation of circular lambda DNA. *Proc. Natl. Acad. Sci. USA* **58**, 240–247.

Green M. R., Maniatis T., and Melton D. A. (1983) Human beta-globin pre-mRNA synthesized *in vitro* is accurately spliced in *Xenopus* oocyte nuclei *Cell* **32**, 681–694.

Grunstein M. and Hogness D. S. (1975) Colony hybridization: A method for the isolation of cloned DNAs that contain a specific gene. *Proc. Natl. Acad Sci. USA* **72**, 3961–3965.

Gubler U. and Hoffman B. J. (1983) A simple and very efficient method for generating cDNA libraries. *Gene* **25**, 263–269.

Hanahan D. (1985) Techniques for transformation of *E. coli*, in *DNA Cloning, vol. 1: A Practical Approach* (Glover, D. M., ed.), pp. 109–135, IRL, Oxford.

Hanahan D. and Meselson M. (1980) Plasmid screening at high colony density. *Gene* **10**, 63–67.

Heidecker G., Messing J., and Gronenborn B. (1980) A versatile primer for sequencing in the M13mp2 cloning system. *Gene* **10**, 69–73.

Helfman D. M., Feramisco J. R., Fiddes J. C., Thomas G. P., and Hughes S. H. (1983) Identification of clones that encode chicken tropomyosin by direct immunological screening of a cDNA expression library. *Proc. Natl. Acad Sci. USA* **80,** 31–35.

Helfman D. M., Feramisco J. R., Fiddes J. C., Thomas G. P., and Hughes S. H. (1985) Identification and isolation of clones by immunological screening of cDNA expression libraries, in *Genetic Engineering: Principles and Methods, vol. 7* (Setlow J. K. and Hollaender A., eds.), pp. 185–197, Plenum, New York.

Helfman D. M. and Hughes S. H. (1987) Use of antibodies to screen cDNA expression libraries prepared in plasmid vectors, in *Methods in Enzymology, vol. 152: Guide to Molecular Clonning Techniques* (Berger S. L. and Kimmel A. R., eds.), pp. 451–457, Academic Press, New York.

Hindley J. (1983) *DNA Sequencing.* Elsevier Biomedical Press, Amsterdam.

Hohn B. and Murray K. (1977) Packaging recombinant DNA molecules into bacteriophage particles *in vitro. Proc. Natl. Acad. Sci. USA* **74,** 3259–3263.

Huynh T. V., Young R. A., and Davis R. W. (1985) Constructing and screening cDNA libraries in lambda gt10 and lambda gt11, in *DNA Cloning, vol. 1: A Practical Approach* (Glover D. M., ed.), pp. 49–78, IRL Press, Oxford.

Jackson D. A., Symons R. H., and Berg P. (1972) Biochemical methods for inserting new genetic information into DNA of simian virus 40: Circular SV40 DNA molecules containing lambda phage genes and the galactose operon of *Escherichia coli. Proc. Natl. Acad. Sci. USA* **69,** 2904–2909.

Jendrisak J., Young R. A., and Engel J. D. (1987) Cloning cDNA into lambda gt10 and lambda gt11 in *Methods in Enzymology, vol. 152: Guide to Molecular Cloning Techniques* (Berger S. L. and Kimmel A. R., eds.), pp. 359–371, Academic Press, New York.

Kacian D. L., Spiegelman S., Bank A., Terada M., Metafora S., Dow, L., and Marks P. A. (1972) *In vitro* synthesis of DNA components of human genes for globins, *Nature New Biol* **235,** 167–169.

Kato K., Goncalves J. M., Houts G. E., and Bollum F. J. (1967) Deoxynucleotide-polymerizing enzymes of calf thymus gland II. Properties of the terminal deoxynucleotidyltransferase. *J. Biol. Chem.* **242,** 2780–2789.

Kimmel A. R. (1987) Identification and characterization of specific clones: Strategy for confirming the validity of presumptive clones, in *Methods in Enzymology, vol. 152: Guide to Molecular Cloning Techniques* (Berger S. L. and Kimmel A. R., eds.), pp. 507–511, Academic Press, New York.

Krug M. S. and Berger S. L. (1987) First-strand cDNA synthesis primed with oligo(dT), in *Methods in Enzymology, vol. 152: Guide to Molecular Cloning Techniques* (Berger S. L. and Kimmel A. R., eds.), pp. 316–325, Academic Press, New York.

Kurtz D. T. and Nicodemus C. F. (1981) Cloning alpha$_{2mu}$ globulin cDNA using a high efficiency technique for the cloning of trace messenger RNAs. *Gene* **13**, 145–152.

Leis J. P., Berkower I., and Hurwitz J. (1973) Mechanism of action of Ribonuclease H isolated from avian myeloblastosis virus and *Escherichia coli*. *Proc. Natl. Acad. Sci. USA* **70**, 466–470.

Lobban P. E. and Kaiser A. D. (1973) Enzymatic end-to-end joining of DNA molecules. *J. Mol. Biol.* **78**, 453–471.

Mandel M. and Higa A. (1970) Calcium-dependent bacteriophage DNA infection. *J. Mol. Biol.* **53**, 159–162.

Maniatis T., Fritsch E. F., and Sambrook J. (1982) *Molecular cloning: A laboratory manual*. Cold Spring Harbor Laboratory, Cold Spring Harbor, NY.

Maxam A. and Gilbert W. (1977) A new method for sequencing DNA. *Proc. Natl. Acad. Sci USA* **74**, 560–564.

Melton D. A., Krieg P. A., Rebagliati M. R., Maniatis T., Zinn K., and Green M. R. (1984) Efficient *in vitro* synthesis of biologically active RNA and RNA hybridization probes from plasmids containing a bacteriophage SP6 promoter. *Nucl. Acids Res.* **12**, 7035–7056.

Mertz J. E. and Davis R. W. (1972) Cleavage of DNA by RI restriction endonuclease generates cohesive ends. *Proc. Natl. Acad. Sci. USA* **69**, 3370–3374.

Messing J. (1983) New M13 vectors for cloning, in *Methods in Enzymology, vol. 101: Recombinant DNA, part C* (Wu R., Grossman L., and Moldave K., eds.), pp. 20–78, Academic Press, New York.

Messing J., Crea R., and Seeburg P. H. (1981) A system for shotgun DNA sequencing. *Nucl. Acids Res.* **9**, 309–321.

Messing J. and Vieira J. (1982) A new pair of M13 vectors for selecting either DNA strand of double-digest restriction fragments. *Gene* **19**, 269–276.

Mierendorf R. C., Percy C. and Young R. A. (1987) Gene isolation by screening lambda gt 11 libraries with antibodies, in *Methods in Enzymology, vol. 152: Guide to Molecular Cloning Techniques* (Berger S. L. and Kimmel A. R., eds.), pp. 458–469, Academic Press, New York.

Murray N. E. and Murray K. (1974) Manipulation of restriction targets in phage lambda to form receptor chromosomes for DNA fragments. *Nature* **251**, 476–481.

Murray N. (1983) Phage lambda and molecular cloning, in *Lambda II* (Hendrix R. W., Roberts J. W. and Stahl F. W., eds.), pp. 395–431, Cold Spring Harbor Laboratories, Cold Spring Harbor, NY.

Newman S., Kitamura K., and Campagnoni A. T. (1987) Identification of a cDNA coding for a fifth form of myelin basic protein in mouse. *Proc. Natl. Acad. Sci. USA* **84,** 886–890.

Okayama H. and Berg P. (1982) High-efficiency cloning of full-length cDNA. *Mol. Cell. Biol* **2,** 161–170.

Olivera B. M. and Lehman I. R. (1967) Linkage of polynucleotides through phosphodiester bonds by an enzyme from *Escherichia coli*. *Proc. Natl. Acad. Sci. USA* **57,** 1426–1433.

Parnes J. R., Velan B., Felsenfeld A., Ramanathan L., Ferrini U., Appella E., and Seidman J. G. (1981) Mouse beta$_2$-microglobulin cDNA clones: A screening procedure for cDNA clones corresponding to rare mRNAs. *Proc. Natl. Acad. Sci. USA* **78,** 2253–2257.

Paterson B. M. and Roberts B. E. (1981) Structural gene identification utilizing eukaryotic cell-free translational systems, in *Gene Amplification and Analysis* (Chirikjian J. G. and Papas T. S., eds.), pp. 417–437. Elsevier/North Holland, Amsterdam.

Plaisancie H., Alexandre Y., Uzan G., Besmond C., Benarous R., Frain M., Trepat J. S., Dreyfus J.-C., and Kahn A. (1984) Immunological screening of standard cDNA libraries in pBR322 vectors: detection of human fibrinogen and prothrombin cDNA clones. *Anal. Biochem.* **142,** 271–276.

Ross J., Aviv H., Scolnick E., and Leder P. (1972) *In vitro* synthesis of DNA complementary to purified rabbit globin mRNA. *Proc. Natl. Acad. Sci. USA* **69,** 264–268.

Roth H. J., Kronquist K., Pretorius P. J., Crandall B. F., and Campagnoni A. T. (1986) Isolation and characterization of a cDNA coding for a novel human 17.3k myelin basic protein (MBP) variant. *J. Neurosci. Res.* **16,** 227–238.

Rougeon F., Kourilsky P., and Mach B. (1975) Insertion of a rabbit beta-globin gene sequence into an *E. coli* plasmid. *Nucl. Acids Res.* **2,** 2365–2378.

Roychoudhury R., Jay E., and Wu R. (1976) Terminal labeling and addition of homopolymer tracts to duplex DNA fragments by terminal deoxynucleotidyl transferase. *Nucl. Acids Res.* **3,** 101–116.

Sanger F., Nicklen S., and Coulson A. R. (1977) DNA sequencing with chain-terminating inhibitors. *Proc. Natl. Acad. Sci. USA* **74,** 5463–5467.

Sgaramella V. (1972) Enzymatic oligomerization of bacteriophage P22 DNA and of linear simian virus 40 DNA. *Proc. Natl. Acad. Sci. USA* **69**, 3389–3393.
Smith H. O. and Birnstiel M. O. (1976) A simple method for DNA restriction site mapping. *Nucl. Acids Res.* **3**, 2387–2398.
Southern E. M. (1975) Detection of specific sequences among DNA fragments separated by gel electrophoresis. *J. Mol. Biol* **98**, 503–517.
Temin H. M. and Mizutani S. (1970) RNA-dependent DNA polymerase in virions of Rous sarcoma virus. *Nature* **226**, 1211–1213.
Thomas M., Cameron J. R., and Davis R. W. (1974) Viable molecular hybrids of bacteriophage lambda and eukaryotic DNA. *Proc. Natl. Acad. Sci. USA* **71**, 4579–4583.
Verma I. M., Temple G. F., Fan H., and Baltimore D. (1972) In vitro synthesis of DNA complementary to rabbit reticulocyte 10S RNA. *Nature New Biol.* **235**, 163–167.
Vieira J. and Messing J. (1982) The pUC plasmids, an M13mp7-derived system for insertion mutagenesis and sequencing with synthetic universal primers. *Gene* **19**, 259–268.
Weiss B. and Richardson C. C. (1967) Enzymatic breakage and joining of deoxyribonucleic acid. I. Repair of single strand breaks in DNA by an enzyme system from *Escherichia coli* infected with T4 bacteriophage. *Proc. Natl. Acad. Sci. USA* **57**, 1021–1028.
Williams S. A., Slatko B. E., Moran L. S., and DeSimone S. M. (1986) Sequencing in the fast lane: A rapid protocol for [alpha-^{35}S] dATP dideoxy DNA sequencing. *BioTechniques* **4**, 138–147.
Wu R., Wu T., and Ray A. (1987) Adaptors, linkers, and methylation, in *Methods in Enzymology, vol. 152: Guide to Molecular Cloning Techniques* (Berger S. L. and Kimmel A. R., eds.), pp. 343–349, Academic Press, New York.
Yanisch-Perron C., Vieira J., and Messing J. (1985) Improved M13 phage cloning vectors and host strains: nucleotide sequences of the M13mp18 and pUC19 vectors. *Gene* **33**, 103–119.
Young R. A. and Davis R. W. (1983a) Efficient isolation of genes by using antibody probes. *Proc. Natl. Acad. Sci. USA* **80**, 1194–1198.
Young R. A. and Davis R. W. (1983b) Yeast RNA polymerase II genes: Isolation with antibody probes. *Science* **222**, 778–782.
Young R. A. and Davis R. W. (1985) Immunoscreening lambda gt11 recombinant DNA expression libraries, in *Genetic Engineering: Principles and Methods, vol. 7* (Setlow J. K. and Hollaender A., eds.), pp. 29–41, Plenum, New York.
Zimmerman S. B., Little J. W., Oshinsky C. K., and Gellert M. (1967) Enzymatic joining of DNA strands: A novel reaction of diphosphopyridine nucleotide. *Proc. Natl. Acad. Sci. USA* **57**, 1841–1848.

Preparation and Use of Subtractive cDNA Hybridization Probes for cDNA Cloning

Gabriel H. Travis, Robert J. Milner, and J. Gregor Sutcliffe

1. Introduction

Molecular cloning is the process of inserting foreign fragments of DNA into a plasmid or bacteriophage vector that is capable of autonomous replication in a suitable host cell. The resulting recombinant DNA molecules can then be amplified by growth in the host and isolated in pure form. The nucleotide sequence of the inserted portion of the recombinant molecule can shed light on the structure of a particular gene or messenger RNA (mRNA) and provide the primary amino acid sequence of the protein it encodes. Because of the greater accuracy and rapidity of nucleotide sequence analysis over conventional protein sequence analysis, this is now the standard method for elucidating the primary structure of a protein. Recombinant clones may be used to generate probes for monitoring mRNA expression either by Northern blotting, RNase protection, or in situ hybridization, and, thus, the developmental and anatomical distribution of the mRNA may be determined. Recombinant clones can be tailored to produce large amounts of the encoded protein in bacterial or mammalian tissue culture systems. The cloned DNA can be altered by any of a number of in vitro mutagenesis procedures, thus altering the amino acid sequence of the encoded protein. The wild-type and mutant proteins can be studied for their functions in the tissue culture cells that express them, or may be purified for study in vitro. Finally, cloned genes are substrates that can be used for altering gene function in experiments with transgenic animals.

mRNA occupies a central position in gene expression and, therefore, the isolation and characterization of clones of mRNAs is central to the analysis of gene expression. RNA itself cannot be cloned with present technology (nor for most purposes is that a desirable goal). However, when RNA is copied into DNA by the enzyme reverse transcriptase and then the product made double-stranded, the resulting complementary DNA (cDNA) is easy to handle because of its relative chemical stability and can be inserted into a vector and cloned. The collection of clones that results from making cDNA from heterogeneous cellular mRNA and inserting the population of molecules into a cloning vector is known as a cDNA library. This chapter will discuss the isolation of cDNA clones from cDNA libraries with examples especially relevant to the cloning of mRNAs of the mammalian central nervous system. New improvements in a strategy called subtractive hybridization are discussed in depth, since we envision that this technology will be extensively utilized in the next generation of cloning experiments.

2. The Arithmetic of mRNA

2.1. Total RNA

In contemplating a cDNA cloning project, several concepts about mRNA should be kept in mind. Most mRNAs end with a poly(A) tail. The major nonpolyadenylated RNAs are ribosomal RNAs, transfer RNAs and so-called small nuclear RNAs that cumulatively make up over 95% of the cellular RNA mass. There is evidence that brain contains another population of RNA molecules that lack poly(A) tails and are of high sequence complexity. It is thought that these might represent additional mRNA molecules (Chikaraishi, 1979; Chaudhari and Hahn, 1983). An alternative explanation is that most of these so-called poly(A)$^-$ mRNAs actually represent degradation products of long poly(A)$^+$ mRNA molecules lacking their extreme 3' ends (Sutcliffe, 1988).

2.2. Polyadenylated RNA

Poly(A)$^+$ RNA can be separated from total cellular RNA because its poly(A) tail will bind by base pairing to oligo(dT) cellulose (Aviv and Leder, 1972). It represents 1.5–2% of the total cellular RNA mass and consists of an estimated $1-3 \times 10^5$ mRNA molecules per cell. For purposes of discussion, these molecules are commonly grouped as highly abundant (more than 1000 mol/cell), moderately abundant (20–1000 mol/cell), and rare (fewer than 20 mol/cell). Each cell type expresses a few mRNAs in the high abundance class that encode the major protein products of those cells. Moderately abundant mRNAs make up a substantial portion of a cell's mRNA mass. However, rare mRNAs, which may constitute only a small portion of the total mass, make up most of the distinct mRNA species. Because of the extreme cellular specialization within the brain, many mRNAs that might be moderately or even highly abundant in the particular cells that synthesize them, are rare in RNA samples isolated from whole brain or even from dissected brain regions.

2.3. mRNA Complexity

The number of distinct mRNAs expressed by the brain, or, for that matter, any other organ can be estimated from measures of the *nucleotide sequence complexity* of the mRNA population. This can be thought of as the total length achieved if each unique mRNA expressed by a tissue were laid end to end. The nucleotide sequence complexity of mRNA from various organs has been measured both by saturation hybridization and by hybridization kinetics analyses (an issue extensively discussed by Milner et al., 1987; *see references* therein). In the rodent brain, roughly 5% of the genome, or $1–2 \times 10^8$ nucleotides, are expressed as stable cytoplasmic mRNA products. Nonneural tissues express less than half this amount. Approximately 65% of the mRNA sequences expressed in rodent brain are not shared with liver or kidney. Brain nuclear RNA is even more complex than stable cytoplasmic RNA, with about 42% of the nonrepetitive genomic sequences expressed. Stable mRNA is derived from this pool by RNA splicing.

To determine the number of mRNAs expressed in a tissue from its mRNA nucleotide sequence complexity, the lengths of the mRNAs must be known. The average length of total poly(A)$^+$ RNA from most vertebrate tissues, including brain, measured by bulk migration either on sedimentation gradients or in gel electrophoresis experiments, ranges from 1500–2000 nucleotides (Meyuhas and Perry, 1979). However, these measurements are dominated by the highly abundant mRNA species that typically are much shorter than are the rarer species. For the brain, the average size of mRNAs has been measured by blot analysis using a large number of randomly selected cDNA clones of brain mRNAs in each abundance class (Milner and Sutcliffe, 1983). The moderately abundant mRNAs have average lengths of 3500–4000 nucleotides, and the rare mRNAs average about 5000 nucleotides. These values have recently been confirmed in a survey of the sizes of known brain mRNA molecules (Sutcliffe, 1988). A typical brain-specific mRNA may have a 3' untranslated region that is considerably larger than its coding block.

Since most of the mRNAs (complexity 1–2 × 10^8 nucleotides) found in brain mRNA samples are in the rare class, and these have average lengths of 5000 nucleotides, it can be calculated that about 30,000 mRNAs are expressed in the brain from a corresponding set of genes. A substantial portion of these genes, probably more than 30% (Sutcliffe, 1988), use alternative mRNA splicing to generate several related mRNA products from single genes. Such alternative splicing will have a negligible effect on the mRNA complexity measurements, but the number of distinct (albeit related) mRNAs and corresponding proteins expressed in the rodent brain must, therefore, significantly exceed 30,000. Approximately 56% of these brain mRNAs are not expressed detectably in liver or kidney.

From this numerology, it is clear that most brain mRNAs must be in the rare class. These considerations have an important impact on the ease with which clones of particular mRNAs can be isolated.

3. General Principles of Cloning

To isolate a cDNA clone of a particular mRNA of interest from a cDNA library, some information is required to distinguish the desired clone from all of the others in the library. The scope of the required purification for cDNA clones of brain mRNAs is apparent from the above discussion: more than 30,000 distinct mRNAs that range in relative concentrations over more than four orders of magnitude. Thus, the isolation of a clone of a low abundance mRNA represents an effective purification factor of greater than 10,000-fold. Successful cloning is at least a three-step process: (1) establishing a cDNA library from RNA isolated from a particular tissue sample; (2) purifying the clone of interest from that cDNA library; and (3) proving that the purified clone is actually of the desired mRNA. Thus, cloning requires that the mRNA of interest be present in the starting mRNA sample. The type of information one has about the mRNA whose clone must be purified from the library determines what sort of cDNA library to construct, the size of the library that must be screened, how the library will be screened, and how proof of identity will be established.

3.1. cDNA Libraries

The first step in isolating a cDNA clone of a particular mRNA species is the generation of a cDNA library (discussed in the previous chapter). The quality of a cDNA library is defined in terms of the average insert size (measured by determining sizes of the inserts of a few dozen randomly selected library members) and the total number of recombinant clones (determined by the number of clones in the original transfection experiment and the percent of these that contain inserts upon empirical investigation). In a library in which the cDNA synthesis is primed with oligo(dT), most clones contain the 3' end of the mRNA and the insert size is very dependent on the integrity of the starting RNA.

The number of recombinant cDNA clones in a library required for a particular cloning experiment is dependent on the abundance of the mRNA being sought according to Eq. 1, in which N = number of recombinant clones in the library, P = probability of the mRNA species being represented and n = abundance of the mRNA species.

$$N = \ln(1 - P)/\ln(1 - n) \qquad (1)$$

Assuming all mRNAs are cloned with equal efficiency, to be 99% certain that a library contains a clone of a particular mRNA present at 0.01% abundance, a library need only contain 46,000 recombinants, whereas 460,000 recombinants are required for a library to contain a 0.001% abundance mRNA clone at the same confidence level. These values are deceptively low, however, since it is frequently necessary to isolate multiple clones of a given mRNA in order to find one containing a full-length insert, even from a good quality library. Thus libraries of the highest quality are required in order to isolate full-length clones of large, rare mRNAs. Since the majority of brain mRNAs are rare and large, high quality libraries are usually necessary for neurobiological investigations.

A variety of cloning vectors are currently available that can be broadly divided into plasmids and bacteriophage lambda. Collectively, these vectors offer many features, including: clusters of unique restriction enzyme recognition sites permitting directional cloning and simplified subcloning, rapid identification of recombinant clones through insertional inactivation of marker genes, the ability to express cloned cDNAs both in bacterial and eukaryotic cells, the ability to synthesize strand-specific complementary RNA (cRNA) from cloned cDNA sequences, and the ability to select for inserts within a defined size range. In general, libraries of comparable quality can be generated in both plasmid and phage lambda systems. Still a third class of cloning vestors are those derived from bacteriophage M13, which can be isolated in both single- or double-stranded forms. Because of its relatively low transformation efficiency and the instability of inserts greater than 1 kb, M13 is not generally used as a primary cDNA cloning vector.

3.2. Clone Isolation Requires Information

The way one goes about isolating a clone from a cDNA library is determined by the starting information one has in hand. Such starting information may be of one or more of the following four types.

3.2.1. Sequence Information

mRNA molecules and cDNA clones of these molecules are composed of linear arrays of nucleotides. A portion of each mRNA contains the coding region for the protein product of that mRNA and the encoded amino acid sequence of the protein is colinear with the mRNA. The most immediately useful and unambiguous type of information that one might have in hand about a desired clone before initiating a cloning project is the nucleotide sequence of the mRNA. This is not to say that the investigator must know the sequence himself, but only that he possess a DNA fragment of that sequence that can be used as a hybridization probe. An example of this would be using an incomplete cDNA clone (from a previous experiment) to screen a cDNA library for a full-length clone. Another example would be using sequence information derived from the literature to direct synthesis of an oligonucleotide that then could be used as a hybridization probe. The sequence in the probe need not be identical to the target sequence in the library. For example, an mRNA clone from one species can be used to isolate the homologous clone from another species, or a clone of one mRNA can be used to isolate a clone of a different but similar mRNA from the same tissue, as was accomplished for the human visual pigment genes (Nathans et al., 1986). In these experiments, the final step in the hybridization reaction is performed under conditions of reduced stringency such that the stability of imperfectly matched duplexes is increased.

Partial amino acid sequence information can also be used to isolate a cDNA clone. Using the genetic code, a group of nucleotide sequences can be derived from the amino acid sequence and an oligonucleotide probe synthesized. However, since each of the 20 amino acids are encoded by more than one nucleotide triplet codon (except for methionine and tryptophan, which are each encoded by a single codon), the derived nucleotide sequence will be degenerate, usually at every third base. The number of possible sequences in the group is known as the level of degeneracy. In selecting a region of sequence for oligonucleotide probe synthesis, it is helpful to maximize the length of the region while minimizing the level of degeneracy. The degeneracy of an oligonucleotide probe can be minimized by selecting a region of the protein sequence that contains amino acids encoded by fewer codons. Oligonucleotide probes should generally be made complementary to the mRNA, that is, they should correspond to the antisense strand so that they may serve additional functions as probes for Northern RNA blots and as primers for extension by reverse transcriptase.

Once a candidate cDNA clone is isolated from a library with probes designed from amino acid or nucleotide sequence information, confirmation of its identity comes from DNA sequence analysis. Sequence identity should extend beyond the region bracketed by the probe used for the clone's detection. Another control that can give confidence that a *bona fide* clone has been isolated is hybridization to an independent DNA probe that does not overlap the probe used initially to isolate the candidate clone.

3.2.2. Antibody to Protein Product

One of the reasons that recombinant DNA technologies have had such a large impact on biochemistry is the difficulty inherent in protein purification and amino acid sequence determination. Usually, it is easier to generate an antibody to a protein than to obtain even a partial sequence. Many proteins are originally defined by antibodies before information is known about either their structure or function.

Antibodies that recognize a protein can be used to isolate a clone of its mRNA from a cDNA expression library. This is especially useful for proteins that are still at an early stage of characterization. Nothing more need be known about the protein beyond the identification of a band in a polyacrylamide gel. The band can be excised and injected directly into a rabbit to raise an immune sera for clone screeing. Monoclonal antibodies, raised against brain fractions, can also be used for screeing a cDNA expression library.

The lambda gt11 system is the most commonly used cDNA expression vector (Young and Davis, 1983). In this system the cDNA is cloned into a site near the end of the E. *coli* lacZ gene (which encodes β-galactosidase) and, when lacZ expression is induced in bacterial cultures, a fusion protein between β-galactosidase and the product of the cloned mRNA is produced. Generation of the fusion protein is dependent on the protein encoded by the cloned mRNA having the same orientation and reading frame as lacZ. Thus only ⅙ clones containing the desired sequence will express the protein. If the protein being used to raise antibodies undergoes posttranslational modification, such as glycosylation or phosphorylation, the antibodies may not recognize the fusion protein produced in unmodified form by bacterial cells. Also, incomplete cDNA clones that encode proteins truncated at the amino terminus may be missing important epitopes, and, thus, may not be recognized by the antibody probe. Because of problems of cross-reactivity, it is necessary to verify independently that a reactive plaque truly contains the clone of interest. Immunoreactivity with independent antisera is one source of confirmation.

3.2.3. Functional Activity

Knowledge about the function of a protein can also assist in the isolation of a cDNA clone. For example, a cDNA clone of mRNA for glutamate decarboxylase (GAD) was isolated from a cat brain lambda gt11 expression library using an antibody probe. The identity of the mRNA was verified by demonstrating antibody-blockable catalysis of glutamate to release CO_2 by phage-infected bacterial extracts (Kaufman et al., 1986). Thus the GAD retained its enzymatic activity when fused to β-galactosidase. In another example, a cDNA clone was isolated solely on the basis of retained functional activity in a fusion protein produced by transfected *E. coli*. In this case, clones of a calmodulin-binding proteins were sought. Instead of using an antibody probe, a mouse brain lambda

Subtractive cDNA Cloning

gt11 library was probed with calmodulin labeled with ^{125}I. A phage clone was isolated with was later shown to produce a fusion protein that did indeed bind calmodulin (Sikela and Hahn, 1987).

An entirely different approach, also employing known functional characteristics of the protein, was used to isolate a cDNA clone of the bovine receptor for substance-K. A stomach cDNA library was constructed in a vector that contained the SP6 bacteriophage promoter upstream from the cDNA insert. Transcription in vitro with SP6 RNA polymerase generated sense-strand cRNA. RNA synthesized from pools of clones was injected into Xenopus oocytes that were then screened for expression of the substance-K receptor by detection of an electrophysiological response to exogenously applied substance-K (Masu et al., 1987). Positive pools were then fractionated to isolate single clones. This approach exploited the ability of oocytes to translate foreign mRNAs and properly modify and incorporate the translation products correctly into their cell membranes.

3.2.4. Physical Information

In some instances, either by design or necessity, the only information existing about an mRNA is the anatomical or developmental distribution of its product. In this situation, difference methods are available for identifying clones of mRNAs expressed in one tissue but absent from (or significantly reduced in) another tissue. Such differential cloning approaches, which will be discussed in technical detail below, have been used to isolate clones of identified mRNAs with known tissue distributions, such as the T-cell receptor (Hedrick et al., 1984) and the fourth complement component (Ogata et al., 1983), as well as clones of previously unknown mRNAs that are interesting on the basis of their tissue distributions, such as, mRNA increased in brain during scrapie infection (Wietgrefe et al., 1985) and an mRNA expressed in the rodent brain at embryonic day 16 at much greater levels than in adult (Miller et al., 1987). Thus, differential cloning does not require structural information about the mRNA or its encoded protein, since the cloning criteria are defined only by the biological problem being addressed. For abundant mRNAs, straightforward differential screening techniques exist. However, because of the high molecular complexity of brain mRNA, and, hence, the requirement of isolating clones of rare mRNAs, the more sophisticated technique of subtractive hybridization is necessary for most neurobiological investigations.

4. Clones of Tissue-Specific mRNAs

4.1. Brute-Force Isolation

Conceptually, the simplest approach to isolate clones of mRNAs present in hypothetical tissue A but absent from hypothetical tissue B would be to select randomly cDNAs from a library prepared from A and then to perform Northern blot analysis on each clone using RNA from A and B, in order to define which mRNA was present in A but undetectable in B. This approach is laborious under the best of circumstances, but its applicability depends on the degree of relatedness of the tissues A and B. For example, when this strategy was applied to identify clones of mRNAs expressed in brain but not liver or kidney, at least 30% of randomly selected brain cDNA clones were positive (Milner and Sutcliffe, 1983). However, the closely related tissue samples, such as different regions within brain, the strategy would be particularly ill-suited given the low percentage of mRNAs with regionally heterogeneous distributions (Travis et al., 1987).

4.2. Plus-Minus Screening

4.2.1. Procedure

A much more efficient differential cloning strategy is differential (plus-minus) colony hybridization (Dworkin and Dawid, 1980). According to this approach, a cDNA library is first prepared from tissue A. Colonies are plated onto two identical sets of filters and their immobilized DNA is probed in parallel with radiolabeled cDNA from A and B. Clones that give a hybridization signal with the cDNA probe from A but not from B are isolated for further analysis.

4.2.2. Applications

Differential colony hybridization can be performed with libraries prepared in either plasmid or phage (differential plaque hybridization). Phage offer the practical advantage of reduced background on the autoradiograph caused by less bacterial debris, since a phage plaque represents an area of decreased density on a lawn of cells. They have the disadvantage that inserts usually must be subcloned before detailed analysis. A new generation of chimaeric phage-plasmid vectors (such as, Stratagene's Lambda-Zap) may circumvent this disadvantage. Differential screening can be performed in both low or high density formats. In a high density screen the library is spread at approximately 10^4 clones/plate, and two replica-lifts onto nitrocellulose or nylon filters are taken from each plate for probing. This procedure offers the advantage of allowing analysis of a large number of clones with minimal effort, but has the disadvantage of frequently giving ambiguous results, especially with clones representing lower abundance mRNAs. In a low density colony screen, individual clones are picked into a defined grid at a density of fewer than 10^3/plate. In this case, each clone has a discrete address, and it is much easier to distinguish nonspecific background from signal. The low density screen has the disadvantage of being more labor-intensive since each colony must be individually picked.

4.2.3. Sensitivity

Because complex (uncloned) cDNA is used as a probe in this type of experiment, the signal intensity for each colony is proportional to the abundance of the mRNA represented by that clone. The hybridization signal from clones representing rarer mRNAs may be so weak that they cannot be distinguished from background. Based on early mixed probe reconstitution experiments, an mRNA of 0.06% abundance was found to be the rarest mRNA whose clone was detectable with a mixed cDNA probe (Dworkin and Dawid, 1980). (Improvements in the methods of probe preparation, Section 5.2.2.7., significantly lower this level). Since much of the brain RNA mass is composed of mRNA species with an abundance below this value, the recovery of brain clones with a differential distribution is far from complete (see subtractive methods, 5.2. below). Also, because mRNAs cloned by differential colony hybridization tend to be from the highest abundance class, it is very common to isolate multiple clones of the same sequence. Northern blot analysis of cDNAs derived from a differential screen provides information that permits the distinction of redundant isolates from unique clones, in addition to confirming the differential distribution and providing an abundance estimate of the cloned mRNA.

5. Subtractive Hybridization

5.1. Standard Procedure

In order to isolate clones of rare, differentially-expressed mRNAs, the probe must be modified so that a greater percentage of the total radioactivity is present in molecules representing those sequences of interest. This increases the ratio of signal-to-background on the autoradiograph that determines the minimum abundance an mRNA must have in the starting sample for its cDNA clone to be detectable. A cDNA probe can be purified in this way through the process of subtractive hybridization. This process involves hybridizing radiolabelled cDNA derived from tissue A to an excess of mRNA from tissue B in a small volume (5–10 µL) at 65°C for 12–24 h. The sequences in common between the two populations will form cDNA-mRNA hybrids. After the hybridization reaction, single-stranded (unhybridized) cDNA (enriched for sequences present in A but absent from B) is separated from double-stranded DNA-RNA hybrids (containing sequences in common between A and B) by hydroxyapatite chromatography (Timberlake, 1980). The subtracted cDNA is then used as a probe in a conventional high density colony screen. Alternatively, this procedure can be used to prepare subtracted cDNA for cloning into a library (Hedrick et al., 1984). The only benefit realized by cloning subtracted cDNA, however, is that a smaller number of colonies need be screened, which seldom represents a significant benefit in itself and affords no improvement in the detection of rare mRNAs. Furthermore, libraries prepared from subtracted cDNA tend to be of poor quality, since single-stranded DNA is degraded on hydroxyapatite.

5.1.1. Reaction Parameters

The extent to which a subtracted probe can be enriched with sequences representing differentially expressed mRNAs is dependent on two parameters: the completeness of the hybridization reaction and the efficiency of separating unhybridized from hybridized cDNA. Many variables, including nucleic acid concentration, reaction time, temperature, ionic strength, pH, sequence complexity and fragment size, affect the rate of nucleic acid hybridization in solution. The most significant variables from the point of this discussion are the concentration of nucleic acids participating in the hybridization and the duration of the reaction. The product of the molar nucleotide concentration in the RNA driver and the reaction time can be used to quantitate the hybridization kinetics. This value is termed R_0t, and is expressed in mole seconds per liter. In subtractive hybridization, the reaction is driven by a large excess of RNA (20–50-fold). Theoretically, the change in concentration of unhybridized RNA during the course of the reaction is insignificant. For this reason, subtractive hybridization displays pseudo first-order kinetics (Hood et al., 1975). The relationship between the ratio of starting (D_0) to final unhybridized (D) DNA concentrations and R_0t can be defined by Eq. (2)

$$D/D_0 = e^{-KR_0t} \qquad (2)$$

in which K is an empirically derived rate constant related to the complexity of the RNA (estimated at $6 \times 10^{-3} M^{-1} \cdot s^{-1}$ for mammalian brain). The maximum theoretical R_0t value obtainable in a subtractive hybridization reaction (18 h at 4 mg/mL) is approximately $1,000 M \cdot$ sec, which results in a D/D_0 value of 0.25%. This means that greater than 99% of the cDNA that shares sequence complementarity with the RNA driver will be driven into hybrids. However, because of the exponential relationship between D/D_0 and R_0t, any reduction in R_0t causes a large increase in the fraction of the starting cDNA that remains unhybridized.

5.1.2. Limitations

Although RNA-driven subtractive hybridization, as outlined above, can theoretically permit the cloning of rare, differentially expressed mRNAs, this strategy in practice has a number of serious disadvantages. First, a large excess of RNA driver over probe (20–50-fold) is required for quantitative subtraction. In order to isolate cDNA clones representing rare mRNAs, at least one secondary screen is necessary. Since the cDNA probe is typically prepared from 2 μg of poly(A)$^+$ RNA from tissue A in order to prepare sufficient mass of the radioactive probe, this means that 80–200 μg of poly(A)$^+$ driver RNA from tissue B is required for the subtractive steps. For comparison, the typical recovery of poly(A)$^+$ RNA from one complete adult rat brain is 20–40 μg. For applications in which small brain regions are to be compared or in which tissue samples are precious (e.g., clinical material), this quantity of RNA driver is prohibitive. A more serious disadvantage of this procedure relates to the chemical stability of the driver RNA. Even if precautions are taken to protect the RNA from nucleases, significant degradation occurs during the prolonged 65°C hybridization (G. T., unpublished observations), resulting in unstable DNA-RNA hybrids and a significant increase in the ratio of D/D_0. Finally, hydroxyapatite chromatography is a relatively inefficient process for separating single- and double-stranded DNA. Hydroxyapatite, which contains calcium phosphate, has an affinity for the phosphate backbone of nucleic acids. Because double-stranded molecules have two phosphate backbones, hydroxyapatite has higher affinity for these than for single-stranded molecules. Bound nucleic acids are eluted from the column by increasing the concentration of sodium phosphate buffer pH 6.5 (PB) or increasing the column temperature. Isolation of single-stranded DNA is normally effected by loading the column at a temperature and PB concentration at which single- and double-stranded molecules are differentially retained. However, the affinity of nucleic acids for hydroxyapatite also varies with their molecular weight. Low molecular weight fragments, even if double-stranded, have a relatively low affinity for hydroxyapatite. Since a sizable proportion of cDNA molecules in a probe mixture are short, there is unavoidable contamination of the single-stranded peak fraction with low molecular weight DNA-RNA hybrids. This contributes significantly to the false positive rate of selected clones. On the other hand, secondary structure in single-stranded cDNA probe molecules may confer double-stranded character that can result in their aberrant retention on hydroxyapatite. Experience has shown that this method gives a high level of false positive clones and a far from quantitative yield (see Travis et al., 1987; Travis and Sutcliffe, 1988).

5.2. Phenol Emulsion-Enhanced DNA-Driven Subtractive Hybridization

5.2.1. Modifications

Several recent improvements on the procedure of subtractive hybridization have circumvented most of these disadvantages (Travis and Sutcliffe, 1988). Rather than using poly(A)$^+$ RNA from tissue B as a subtraction driver, double-stranded plasmid cDNA can be used. This involves generation of a cDNA library from tissue B followed by plasmid amplification. Since a good quality cDNA library can be generated from 2 μg of poly(A)$^+$ RNA, the quantity of RNA from tissue B required for preparing a subtracted probe is thus greately reduced. Furthermore, once cloned the supply of plasmid driver is indefinitely replenishable. Also, because the hybridization driver is DNA rather than RNA, susceptibility to nucleases and low chemical stability are much less of a problem. Finally, since the hybridization reaction involves DNA with DNA, the kinetics can be significantly increased by performing the reaction in a phenol emulsion (Kohne et al., 1977). To be accurate, this reaction involves competitive hybridization rather than subtractive hybridization since the probe must compete with the anti-sense strand of the driver. Assuming that the reaction is driven to completion, the percent of common sequences in the probe that remain unhybridized is inversely proportional to the excess of driver over probe. In a 100-fold excess of driver over probe (an easily attained value), 99% of the cDNA probe not unique to sample A will be driven into double stranded hybrids, whereas 1% will remain single-stranded owing to competition with the same strand of the driver.

As the first stage of a molecular analysis of the primate cerebral cortex, we employed this modified subtractive hybridization procedure to generate a collection of clones of mRNAs expressed in cerebral cortex of adult Cynomolgus monkey but absent from cerebellum. We have isolated clones of several rare (0.001%), differentially expressed mRNAs, and have recovered clones of at least one moderately abundant (0.05%), differentially expressed mRNA nearly quantitatively from the starting library (Travis and Sutcliffe, 1988). In order to demonstrate the methodology involved in ex eculing a subtractive cloning experiment, and to provide an example of the sort of results one obtains, a detailed account of the above-cited study follows with comments on the critical steps. A flow chart indicating the order of steps is shown in Fig. 1.

5.2.2. Protocol

5.2.2.1. PREPARATION OF RNA. The first step in any cloning experiment is the preparation of poly(A)$^+$ RNA from the tissues of interest. RNA must be isolated with minimal degradation and free of activities that might inhibit later reactions. No cDNA library can be of higher quality than the mRNA used to generate it. We routinely have success with the cytoplasmic RNA preparation method of Schibler et al. (1980) for fresh tissues, and the total cellular RNA preparation method of Chirgwin et al. (1979) for frozen material such as that from human autopsy. Enrichment for

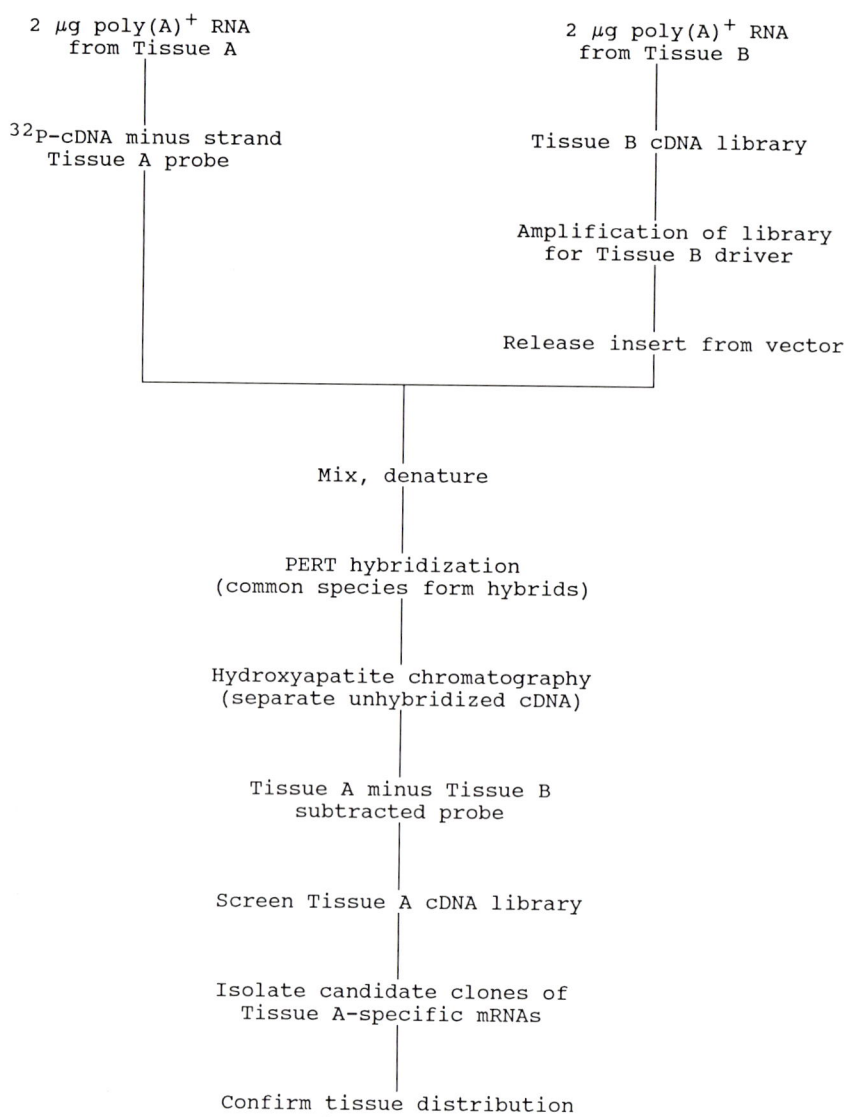

Fig. 1. Flow diagram of PERT-enhanced DNA-driven subtractive hybridization strategy for isolating clones of tissue-specific mRNAs.

Subtractive cDNA Cloning

poly(A)$^+$ RNA is performed on oligo-dT cellulose (Aviv and Leder, 1972) using the conditions described by Maniatis et al. (1982).

5.2.2.2. CDNA CLONING AND AMPLIFICATION. In our experiments, monkey cerebral cortex and cerebellum cDNA libraries were constructed in the plasmid vector pGEM4 using a vector-primed cloning procedure that enriches for full-length clones (Milner et al., 1985) beginning with 2 µg of poly(A)$^+$ RNA from each tissue. The pGEM4 vector was chosen because cRNA transcripts of the insert can easily be generated for later use as *in situ* hybridization probes. The two libraries were used to transform *E. coli* strain MC1061. In this experiment, each library was estimated to contain 10^6 clones, determined by plating an aliquot of each transformation mixture directly onto an ampicillin selection plate prior to amplification. Mini plasmid preparations were performed on 16 colonies randomly selected from the plate, and these were analyzed by restriction digestion and agarose gel electrophoresis. Fourteen out of 16 digested plasmids had inserts that varied in size from a few hundred bases to 1.5 kb.

The libraries were amplified by growing the transformed cells in liquid media under ampicillin selection (Okayama and Berg, 1983). The concern is frequently voiced that amplifying a cDNA library will result in a skewed representation of clones relative to the template RNA. The basis for this objection is that certain cDNA sequences may confer a growth disadvantage (or advantage) to the host bacteria cell. There are documented cases that show that this can occur, but it is impossible to predict whether a particular sequence might be over- or under-represented in an amplified library. In the present discussion, only sequences that are under-represented are a concern, and under-representation could occur in either the A or the B library. For library A, any cDNA sequence that causes the bacteria cell to grow significantly more slowly in liquid culture during an amplification, will also prevent the development of a visible colony or plaque during overnight growth on a plate in an unamplified library. Thus, although it is probably true that clones of some mRNAs are difficult to isolate under any conditions, it is unlikely that many cDNAs can be isolated from an unamplified but not an amplified library. One effect of amplification is that plasmids bearing very large inserts will tend to be present at a lower copy number and thus will be under-represented in the library. However, library amplification permits the DNA to be size selected by electrophoresis on agarose, circumventing the problem of a bias toward smaller inserts and permitting the analysis of clones with any desired insert size range. If a particular cDNA is under-represented in library B, this could give rise to false positives when the B library is used as a hybridization driver. Thus, each clone picked as a positive with a subtractive probe must be considered as a candidate: proof that the corresponding mRNA is truly differentially expressed awaits direct measurement of the mRNA in the two tissues.

According to the subtractive strategy described here, it is not necessary that the library used for clone isolation be amplified or even constructed in plasmid. However, it is necessary that the library that functions as a subtractive hybridization driver be amplified in plasmid, since 120 μg of cDNA must be prepared (see below). We find it convenient to prepare and amplify both the A and B libraries in parallel, thus ensuring a permanent stock of DNA for subsequent experiments.

In our experiment, the cerebral cortex library was size-selected by electrophoretically separating 5 μg of the amplified, supercoiled library DNA on agarose and isolating plasmids containing inserts between 0.5–6 kbp. The purpose of this size selection was to eliminate clones with very small or no inserts (which potentially have a growth advantage and might be over-represented) without significantly reducing the clonal complexity of the library. Approximately sixty thousand colonies from this size-selected cortex library were spread on six 150 mm LB plates containing 100 μg/mL ampicillin. Filter-lifts, prehybridization, and hybridization were performed following standard high density colony hybridization procedures (Grunstein and Wallis, 1979) with the exception of probe-preparation (see below, 5.2.2.4.). The master plates were stored at 4°C for up to two weeks.

5.2.2.3. PREPARATION OF CEREBELLUM cDNA DRIVER. To make driver cDNA, 120 μg of cerebellum amplified library DNA was digested with restriction enzymes PstI and Eco RI to release inserts. This step generated sufficient hybridization driver for two PERT (phenol emulsion reassociation technique) reactions. Since aproximately 75% of the DNA mass in this sample is plasmid vector, which cannot take part in the hybridization but does lower the efficiency of the PERT reaction, this procedure was not optimized kinetically. In subsequent experiments when purified insert has been used, there was no significant improvement. This is probably because of the fact that the C_0t values obtained, even in a nonoptimized reaction, far exceed the ability of hydroxyapatite to discriminate single- and double-stranded molecules. However, it is necessary to linearize the plasmid DNA, since supercoiled molecules would rapidly reanneal in the PERT reaction, and, thus, would be unavailable for subtraction.

5.2.2.4. PREPARATION OF cDNA PROBE FOR SUBTRACTION. A ^{32}P-labeled cDNA probe was prepared from 2 µg of monkey cortex poly(A)$^+$ RNA by the following procedure.

1. 1 mCi of [α^{32}P]-dCTP (3,000 Ci/mmole) was placed in a sterile, siliconized screwcap 1.5 mL tube and evaporated to dryness in a Speed Vac (Savant).
2. We then added at room temperature:
   15.5 µL dH$_2$0
   10 µL of 5 X RT buffer (1 X = 50 mM Tris pH 8.3; 8 mM MgCl$_2$; 30 mM KC1)
   10 µL Oligo dT(12–18) 1 mg/mL (Pharmacia)
   5 µL of 10 mM dATP, dGTP and TTP (Pharmacia)
   2.5 µL 20 mM dithiothreitol
   1 µL 30 U/µL human placenta RNase inhibitor (Pharmacia)
   4 µL 0.5 mg/mL monkey cortex poly(A)$^+$ RNA
   2 µL 13.5 U/µL AMV reverse transcriptase (Life Sciences)
3. This reaction mixture was incubated for 0.5 h at 42°C.
4. 5 µL of 10 mM dCTP was added, and the reaction was incubated an additional 0.5 h at 42°C. This chase reaction significantly increased the average size of labeled cDNA molecules, stabilizing the hybrid forms during hydroxylapatite chromatography.
5. 0.5 µL was removed to a tube containing 19.5 µL TE for trichloroacetic acid (TCA) precipitation. TCA precipitable counts were measured by spotting 2 µL onto a GF/C filer (Whatman), precipitating the remaining 18 µL by adding 10 µL 10 mg/mL salmon sperm carrier DNA and 1 mL of 20% TCA, incubating on ice for 15 min, and then collecting the precipitate on a 0.45 µm nitrocellulose filter (Schleicher and Schuell) in a Millipore filter apparatus. Both filters were counted in a liquid scintillation counter. Incorporated into cDNA was 42% of the [α^{32}P]-dCTP giving an estimated probe mass of 160 ng.

5.2.2.5. SUBTRACTIVE HYBRIDIZATION REACTION. It has been observed that the rate at which DNA hybridizes with itself can be increased many thousand-fold by performing the reaction in a phenol emulsion (Kohne et al., 1977). The precise mechanism for this enhancement is not known, but it is probably the result of surface effects between the phenol and aqueous phases. For poorly understood reasons, PERT has a minimal effect on the rate of DNA-RNA hybridizations (Kohne et al., 1977).

1. 60 μg of the restriction enzyme-digested cerebellum cDNA driver was added to the ^{32}P-labeled cortex cDNA probe, and the mixture was made 0.3 M in sodium acetate and precipitated with 2.5 volumes of ethanol. The pellet was rinsed with 80% ethanol and lyophilized to dryness. (Note: the supernatants are highly radioactive, hence, adequate care should be taken).
2. The pellet was dissoved in 10 mM Tris pH 8.0, 1 mM EDTA.
3. The DNA mixture was denatured by heating to 100°C for 5 min.
4. Sodium thiocyanate (2M final concentration) and phenol (10% final concentration) were added to give a final volume of 800 μL.
5. Hybridization was performed at 25°C with continuous agitation on a vortex mixer for 4 d. [More recently, better results have been obtained with 1-d hybridizations, which minimize radiolytic decay of the probe.] This can be accomplished by fixing the reaction vessel in the end of a short piece of plastic laboratory tubing (to provide flexibility) and clamping the tubing firmly above the vortex mixer. The bottom of the reaction vessel should be reinforced (with a small piece of tubing), since prolonged friction will wear through. During the first hour of operation, the apparatus should be monitored, since the strength of the mixing will change as the motor warms up.
6. After hybridization, the mixture was extracted with chloroform, the aqueous phase precipitated with 2 mL of ethanol, and then the pellet dissolved in 1 mL

Subtractive cDNA Cloning

of 50 mM sodium phosphate buffer, pH 6.5 (PB) for hydroxyapatite chromatography.

We estimate that a C_0t value of 20,000–80,000M/s is achieved under these conditions.

5.2.2.6. HYDROXYAPATITE CHROMATOGRAPHY. To improve the efficiency of separating single- from double-stranded DNA, and to lower the colony background, we have modified the standard protocol by adding a 50 mM PB elution step, which we have shown removes low molecular weight DNA fragments (Travis and Sutcliffe, 1988).

1. A 3 mL bed volume of hydroxyapatite (DNA Grade Bio-Gel HTP, Bio Rad) equilibrated with 50 mM PB, 0.2% sodium dodecyl sulfate (SDS), was poured into a jacketed column (Bio Rad). The column and buffers were maintained at 60°C throughout the separation.
2. The hybridization mixture in 1 mL 50 mM PB was loaded onto the column. The column was washed with 9 mL 50 mM PB, 0.2% SDS and 10 mL of eluate was collected at a flowrate of 1 mL/min. This low molecular weight fraction typically contains 10–12% of the total radioactive counts. When visualized by gel electrophoresis and autoradiography, a smear of low molecular weight fragments (<500bp) was observed.
3. The single-stranded fraction was eluted with 10 mL of 140 mM PB, 0.2% SDS. One mL samples were collected.
4. The remaining double-stranded material was eluted with 10 mL 400 mM PB, 0.2% SDS.

The single-stranded fraction contained 5% of the total eluted counts. The three 1 mL samples which contained most of the counts within the 140 mM PB fraction were added directly to the high density colony hybridization mixture. Six-hundred-sixty colonies of approximately 60,000 in the cortex library (approx 1%) gave hybridization signals with this subtracted cortex probe (example in Fig. 2). These were picked (see below) to replica filters.

5.2.2.7. DIFFERENTIAL COLONY HYBRIDIZATION. We next performed a differential colony hybridization screen on these subtraction-selected clones using unsubtracted cortex and cerebellum

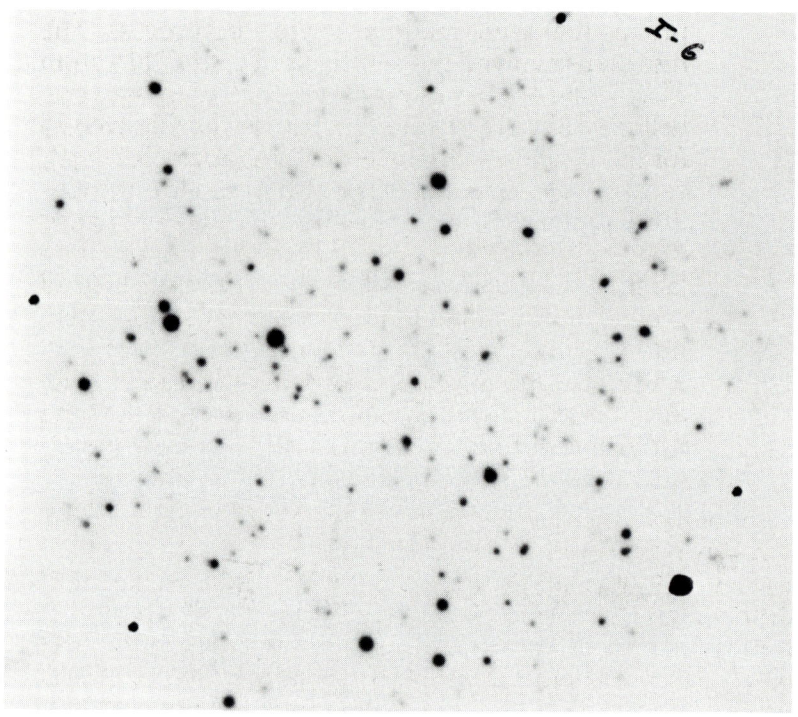

Fig. 2. Representative autoradiograph from high density subtractive screen. Approximately 10,000 monkey cortex cDNA clones per filter were probed with a monkey cortex-minus-cerebellum cDNA preparation. Five of the peripheral spots were hand drawn marks for filter orientation.

cDNA probes. This was done to detect clones of moderately abundant mRNAs present in cortex but absent from cerebellum.

1. Each of these 660 clones was individually picked and placed into a grid composed of ten 96-well microtiter dishes containing 160 μL of LB freezing media (Gergen et al., 1979) in each well. This allowed the selected colonies to be frozen and stored indefinitely. A duplicate set of these colonies were innoculated onto Biotrans membranes overlying LB plates using a multiprong device.
2. Differential colony hybridization was performed using published procedures (Dworkin and Dawid, 1980) except for the preparation of the unsubtracted

probes that were made in a two step process. The first step involved the synthesis of cold cDNA from 2 µg poly(A)$^+$ monkey cortex and monkey cerebellum RNA. The same procedure was followed as for the synthesis of labelled cDNA (above 5.2.2.4.), except that unlabeled dCTP (1 mM final concentration) replaced the [α^{32}P]-dCTP. Thus, unlabeled first strand cDNA was synthesized. A separate labeling reaction was performed to generate labeled second strand using random hexamer oligonucleotide priming (Feinberg and Vogelstein, 1983) with 100 µCi of [α^{32}P]-dCTP. This two-step procedure appears to increase significantly the sensitivity and reduce the number of false positives on the differential screen (G. T., unpublished observations).

A total of one hundred-sixty colonies (24%) gave a signal with the cortex probe but not the cerebellum probe in this assay. These represented candidate cDNA clones of "cortex-specific" mRNAs. Additionally, 37 colonies (6%) gave no hybridization signal with either the cortex or cerebellum probes. The remainder of clones gave signals with both cortex and cerebellum probes, representing sequences incompletely subtracted during the hybridization reaction or human error in picking colonies from high density plates. (In more recent experiments, in which the density of the colonies in the initial screen was reduced, fewer false positives were seen). Figure 3 shows a sample set of filters from this differential screen.

Ten of these 160 candidate "cortex-specific" cDNAs were selected at random and used to probe Northern blots containing poly(A)$^+$ RNA from monkey cortex, cerebellum, and liver, as well as from frontal cortex of Alzheimer's disease and normal human brain autopsy specimens. All ten clones detected RNAs that were either specific to cortex at the sensitivity of Northern blot analysis or significantly enriched in cortex over cerebellum.

We reprobed the same low density grid with clone pMC1H8a, a cDNA clone representing a monkey "cortex-specific" mRNA with an estimated abundance of 0.05% isolated in a previous screen. Clones containing sequences homologous to pMC1H8a were detected in 59 of the 660 subtractive-selected colonies. When the original high-density filters from the cortex cDNA library were reprobed with pMC1H8a, it was found to hybridize to 60 of the 60,000 unselected colonies. From this we concluded that the recov-

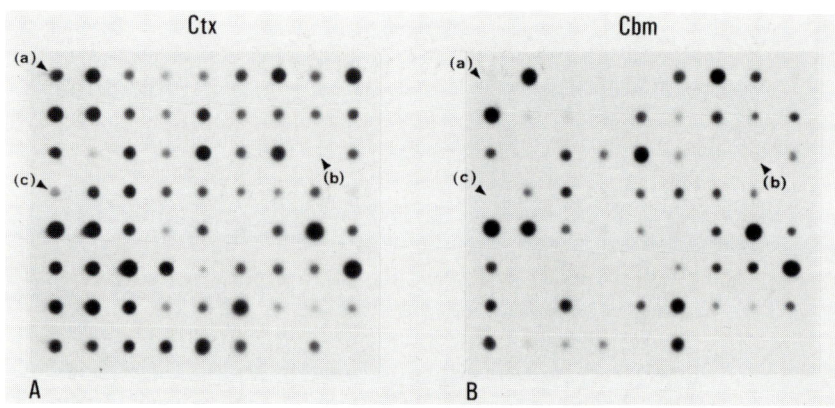

Fig. 3. Representative set of differential colony screen autoradiographs. Clones positive with the subtracted probe were picked into a grid and replica filters produced. Filters were probed pairwise with unsubtracted monkey cortex (Ctx) and cerebellum (Cbm) ^{32}P-labeled cDNA (5.2.2.7.). Clones pMC2A1 (a) and pMC2D1 (c) each gave hybridization signals with the unsubtracted cortex but not the unsubtracted cerebellum probes, and later Northern blot analysis showed that each hybridized to an mRNA highly enriched in cortex. Clone pMC2C8 (b) gave no detectable signal with either of the unsubtracted probes (see Figs. 3 and 4).

ery of pMC1H8a-homologous clones with the subtracted probe from the starting library was nearly quantitative.

5.2.2.8. SECONDARY SUBTRACTIVE SCREEN. The 37 colonies that gave no signal with either of the unsubtracted probes potentially included clones of rare cortex-specific mRNAs. To detect such clones, the 37 colonies were placed into a separate grid and reprobed with another preparation (see 5.2.2.4.) of cortex-minus-cerebellum subtracted cDNA. Nine of these colonies gave a clear signal with this second subtracted probe (Fig. 4). The remaining 28 clones, negative with the second subtracted probe, probably arose from unavoidable errors in picking colonies from the high density screen. The nine positive clones represented candidate rare "cortex-specific" cDNAs and were used for Northern blot analysis. Three of these detected mRNAs that were "cortex specific" at the sensitivity of the assay (Fig. 5).

The absolute abundances for two of these mRNAs were determined by comparing their autoradiograph signal densities when probed with nick-translated whole plasmid with the signal density from 1 pg of plasmid DNA on the adjacent lane in a

Subtractive cDNA Cloning

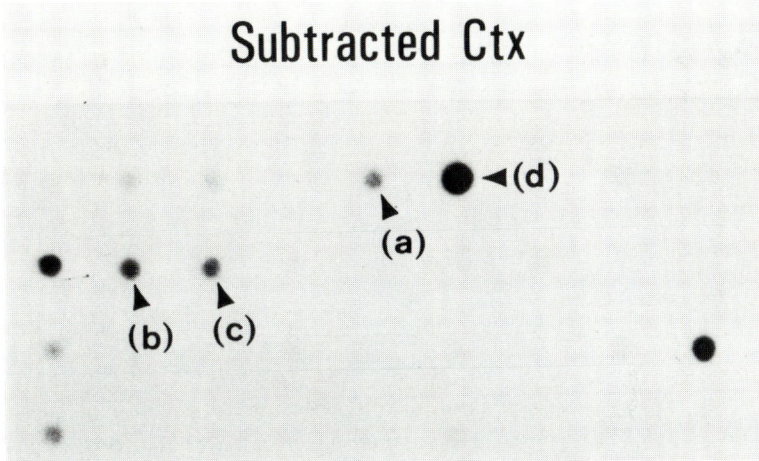

Fig. 4. Subtractive probe colony autoradiograph of 37 clones previously negative with both unsubtracted cortex and cerebellum cDNAs. Nine colonies gave a clear hybridization signal in this experiment when reprobed with cortex-minus-cerebellum cDNA preparation. Clone pMC2C8 (a), pMC4G8 (b) and pMC5B3 (c). Clone pMC2D6 (d) gave a signal with the unsubtracted monkey cortex probe and was included as a positive control.

Northern blotting experiment. Assuming equal transfer of similar-sized RNA and DNA fragments, and equal stability of the DNA-DNA and DNA-RNA hybrids under our stringency-wash conditions, an abundance estimate in monkey cortex poly(A)$^+$ RNA of 0.001% was obtained for both of these mRNA species. For these calculations we took into account the relative lengths of the DNA and RNA targets, and that both strands of the DNA were available for hybridization.

5.2.3. Quantitative Assessment

The use of subtractive hybridization in this type of experiment has advantages over plus-minus screening. First, it is more sensitive, permitting the isolation of clones of rare, differentially expressed mRNAs. Additionally, for plus-minus screening to be effective, it must be performed at low density, which means manually picking colonies into grids. Screening 60,000 clones, which was accomplished easily here by subtraction, would be a formidable challenge by plus-minus screening, and would not be expected to yield clones of the rarer mRNAs.

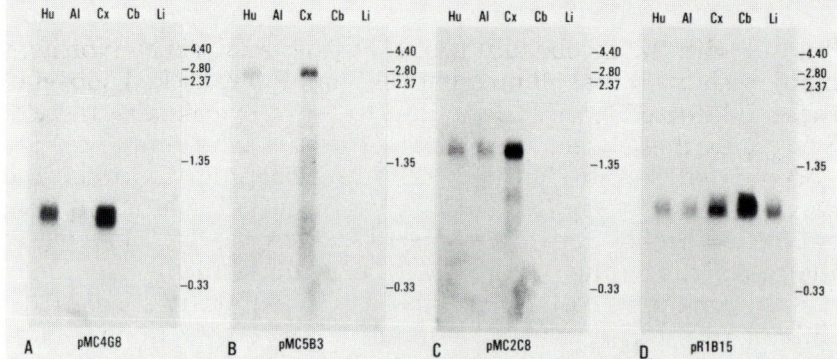

Fig. 5. Northern blot analysis of three clones (cf., Fig. 3) representing rare mRNAs selected with a cortex-minus-cerebellum subtracted cDNA probe. RNA samples (2 μg poly(A)$^+$) included normal human frontal cortex (Hu), Alzheimer's disease frontal cortex (Al), monkey cerebral cortex (Cx), monkey cerebellum (Cbm) and monkey liver (Li). Blots were probed with a cDNA clone of rat cyclophilin (p1B15), an ubiquitously expressed mRNA (D), as a control for gel loading and RNA integrity. The RNA size standard positions (kb) are indicated.

From 60,000 cortex cDNA clones originally screened with a probe produced by PERT-enhanced, DNA-driven subtractive hybridization, as described above, a collection of 660 clones resulted. The efficiency of the subtracted probe in detecting cortex-specific mRNAs was measured directly. Judging by the isolation of 59 of the possible 60 pMC1H8a-homologous clones, the recovery is nearly quantitative, at least for clones of moderately abundant mRNAs. Thus, use of a subtracted probe in the first step of this differential screening experiment afforded approximately a 100-fold enrichment of the class of mRNA clone being sought. This allowed us to pick 100-fold fewer colonies (660 vs 60,000), while discarding hardly any cortex-specific mRNA clones, at least in the abundance class of pMC1H8a (0.05%). This is a compelling reason to use subtractive hybridization, even if low-abundance mRNAs are not being pursued.

The second step in the procedure was to examine the set of 660 clones by plus-minus hybridization. About 70% of the clones hybridized with both cortex and cerebellum probes, and, thus, turned out to be false positives. Many of these were probably a result of human error in colony picking. This number can probably be reduced by performing the initial colony screen at a lower density. About one-quarter of clones hybridized with the cortex but not with the cerebellum probe. Further analysis by Northern blotting showed that all members of a randomly picked subset of these hybridized to mRNAs that were cortex specific or enriched. Several of these were present at below 0.06% abundance, the documented sensitivity of plus-minus screening (Dworkin and Dawid, 1980), suggesting that the modified method of unsubtracted probe preparation described above (5.2.2.7.) increases the sensitivity of the plus-minus screen substantially.

Within the set of 37 clones picked in the subtractive screen that did not give signals in the plus-minus screen were 3 clones of rare (approximately 0.001%) mRNAs highly enriched in cortex. One of these clones (pMC4G8, in Fig. 5) was shown by sequence analysis to correspond to the mRNA for preprosomatostatin, a protein known to be rare in cerebral cortex and absent from cerebellum. These 3 clones could not have been isolated using normal plus-minus screening because of the rarity of their mRNAs.

The straightforward series of steps outlined above led to the isolation of 163 clones of the desired mRNAs from a starting pool of 60,000 clones. The mRNAs ranged in abundance from 0.05 to 0.001%. After the first step, 70% of the clones were false positives. Given the degree of purification and near quantitative recovery in this first step, this rate of false positives seems acceptable for most applications. The size of the collection at this stage is easy to handle.

The one step in this subtractive cloning strategy that is still far from optimal is the separation of single- and double-stranded cDNAs on hydroxyapatite. In order to isolate clones of rare, differentially expressed mRNAs quantitatively, it will probably be necessary to reduce the proportion of the single-stranded fraction to below 1% of the total incorporation. One approach, which has offered encouraging results in preliminary experiments, is biotinylating the driver cDNA using terminal deoxynucleotide transferase, and then removing hybrid molecules after the PERT reaction by incubation of the DNA with Streptavidin agarose (G. T., unpublished observations). This step may completely replace hydroxyapatite chromatography, or may be used in series with it since the physical basis of the two separations are unrelated.

6. Future Applications

The technology as it currently stands permits the isolation of clones of most differentially expressed, moderately abundant mRNAs and at least some differentially expressed, rare mRNAs. It is both sensitive and selective, producing a manageably small number of false positives that can easily be weeded out in a second screen. It requires only a modest input of tissue. This reduction in the tissue requirement opens the door to a number of biological problems. In principle, the molecular basis that accounts for the differences between any two tissues can be investigated.

Of neurobiological interest, clones of mRNAs present in one small region of the brain, but absent from other small regions, can now be isolated. For example, this approach could be used to search for mRNAs present in one area of cerebral cortex but absent from another cortical area. It should be possible to isolate clones of developmentally regulated mRNAs present in early embryonic tissue by subtracting between developmental stages. The inductive effects of hormones or other pharmacologic agents on small organ systems can be studied. For example, one might look for mRNAs induced in the pituitary after castration. Finally, disease mechanisms can be approached by subtractive cloning of RNAs from normal abnormal tissues. For instance, any genetically determined, focal degenerative disease is a candidate for study. An example would be looking for mRNAs present in the caudate nucleus of normal human brain but absent from Huntington's disease brain.

Acknowledgments

We thank Juan Bernal, Miles Brennan, Stefano Catsicas, Patria Danielson, Douglas Feinstein, Dan Larhammar, and Hans Peter Ottiger for constructive comments, and Linda Elder for preparing the manuscript. This is publication no. 5225-MB from the Research Institute of Scripps Clinic. We acknowledge the National Institutes of Health (NS22347, NS22111, and NS00918) for partial support.

References

Aviv H. and Leder R. (1972) Purification of biologically active globin messenger RNA by chromography on oligothymidylic acid-cellulose. *Proc. Natl. Acad. Sci. USA* **69**, 1408–1412.

Chaudhari N., Hahn W. E. (1983) Genetic expression in the developing brain. *Science* **220**, 924–928.

Chirgwin J. M., Przybyla A. E., MacDonald R. J., and Rutter W. J. (1979) Isolation of biologically active ribonucleic acid from sources enriched in ribonuclease. *Biochemistry* **18**, 5294–5299.

Chikaraishi D. M. (1979) Complexity of cytoplasmic polyadenylated and nonpolyadenylated rat brain ribonucleic acid. *Biochemistry* **18**, 3249–3256.

Dworkin M. B. and Dawid I. B. (1980) Use of a cloned library for the study of abundant poly(A)$^+$ RNA during *Xenopus laevis* development. *Dev. Biol.* **76**, 449–464.

Feinberg A. P. and Vogelstein B. (1983) A technique for radiolabeling DNA restriction endonuclease fragments to high specific activity. *Anal. Biochem.* **132**, 6–13.

Gergen J. P, Stern R. H., and Wensink P. C. (1979) Filter replicas and permanent collections of recombinant DNA plasmids. *Nucl. Acids Res.* **7**, 2115–2136.

Grunstein M. and Wallis J. (1979) *Methods in Enzymology* (Wu R., ed.), vol. 68, pp. 379–389. Academic Press, New York.

Hedrick S. M., Cohen D. I., Nielsen E. A., and Davis M. M. (1984) Isolation of cDNA clones encoding T cell-specific membrane-associated proteins. *Nature* **308**, 149–153.

Hood L. E., Wilson J. H., and Wood W. B. (1975) *Molecular Biology of Eucaryotic Cells. A Problems Approach.* vol. 1, The Benjamin/Cummings Publishing Company, Menlo Park, California.

Kaufman D. L., McGinnis J. F., Krieger N. R., and Tobin A. J. (1986) Brain glutamate decarboxylase cloned in λgt-11: Fusion protein produces δ-aminobutyric acid. *Science* **232**, 1138–1140.

Kohne D. E., Levison S. A., and Byers M. J. (1977) Room temperature method for increasing the rate of DNA reassociation by many thousandfold: The phenol emulsion reassociation technique. *Biochemistry* **16**, 5329–5341.

Maniatis T., Fritsch E. G., and Sambrook J. (1982) *Molecular Cloning.* Cold Spring Harbor, New York: Cold Spring Harbor Laboratories.

Masu Y., Nakayama K., Tamaki H., Harada Y., Kuno M., and Nakanishi S. (1987) cDNA cloning of bovine substance-K receptor through oocyte expression system. *Nature* **329**, 836–838.

Meyuhas O. and Perry R. P. (1979) Relationship between size, stability, and abundance of the messenger RNA of mouse L cells. *Cell* **16**, 139–148.

Miller F. D., Naus C. C. G., Higgins G. A., Bloom F. E., and Milner R. J. (1987) Developmentally regulated rat brain mRNAs: Molecular and anatomical characterization. *J. Neurosci.* **7**, 2433–2444.

Milner R. J. and Sutcliffe J. G. (1983) Gene expression in rat brain. Nucl. Acids Res. **11**, 5497–5520.

Milner R. J., Lai C., Nave K-A., Lenoir D., Ogata J., and Sutcliffe J. G. (1985) Nucleotide sequences of two mRNAs for rat brain myelin proteolipid protein. Cell **42**, 931–939.

Milner R. J., Bloom F. E., and Sutcliffe J. G. (1987) Brain specific genes: Strategies and issues. *Current Topics in Developmental Biology* **21**, 117–150.

Nathans J., Thomas D., and Hogness D. S. (1986) Molecular genetics of human color vision: The genes encoding blue, green, and red pigments. *Science* **232**, 193–202.

Ogata R. T., Shreffler D. C., Sepich D. S., and Lilly S. P. (1983) cDNA clone spanning the α-δ subunit junction in the precursor of the murine fourth complement component (C4). *Proc. Natl. Acad. Sci. USA* **80**, 5061–5065.

Okayama H. and Berg P. (1983) A cDNA cloning vector that permits expression of cDNA inserts in mammalian cells. *Mol. Cell. Biol.* **3**, 280–289.

Schibler K, Tosi M., Pittet A.-C., Fabiani L., and Wellauer P. K. (1980) Tissue-specific expression of mouse α-amylase genes. *J. Mol. Biol.* **142**, 93–116.

Sikela J. M. and Hahn W. E. (1987) Screening an expression library with a ligand probe: Isolation and sequence of a cDNA corresponding to a brain calmodulin-binding protein. *Proc. Natl. Acad. Sci. USA* **84**, 3038–3042.

Sutcliffe J. G. (1988) mRNA in the mammalian central nervous system. *Ann. Rev. Neurosci.* **11**, 157–198.

Timberlake W. E. (1980) Developmental gene regulation in *Aspergillus nidulans*. *Dev. Biol.* **78**, 497–510.

Travis G. H., Naus C. G., Morrison J. H., Bloom F. E., and Sutcliffe J. G. (1987) Subtractive cloning of complementary DNAs and analysis of messenger RNAs with regional heterogeneous distributions in primate cortex. *Neuropharmacology* **26**, 845–854.

Travis G. H. and Sutcliffe J. G. (1988) Phenol emulsion-enhanced DNA-driven subtractive cDNA cloning: Isolation of low abundance monkey cortex-specific mRNAs. *Proc. Natl. Acad. Sci. USA* **85**, 1696–1700.

Wietgrefe S., Zupancic M., Haase A., Chesebro B., Race R., Frey II W., Rustan T., and Friedman R. L. (1985) Cloning of a gene whose expression is increased in scrapie and in senile plaques in human brain. *Science* **230**, 1177–1179.

Young R. A. and Davis R. W. (1983) Efficient isolation of genes by using antibody probes. *Proc. Natl. Acad. Sci. USA* **80**, 1194–1198.

Analysis of Brain-Specific Gene Products

Robert J. Milner and J. Gregor Sutcliffe

1. Introduction

This chapter is concerned with the question, "Now that I've isolated a cDNA clone, what do I do with it?" The discussion is directed particularly at the analysis of clones of mRNAs encoding proteins of unknown identity that have been selected by procedures such as subtractive hybridization (*see* chapter by Travis et al.). In these cases, the question can be rephrased as, "What can I learn about the properties of the putative encoded protein that will give me clues as to its function?" Much can be gained from analyzing the sequence of the putative protein, determining the regional and cellular sites of expression of the protein and its mRNA, and their temporal expression during development. This type of analysis, which will also be of use to those studying clones of identified mRNAs, can often provide sufficient information to generate hypotheses for the function of the protein. We will also discuss procedures for generating antibodies against a protein encoded by a cDNA clone, using synthetic peptides or proteins generated in expression systems as immunogens. These antibodies can be used to demonstrate that the putative protein does indeed exist, to define its biochemical properties, and to test the functional hypotheses experimentally. Ultimately, definition of function may require additional information, and a certain amount of luck. The latter is outside the scope of this chapter.

The cDNA clones to be analyzed may be generated in a variety of ways. For example, in our laboratories we have selected cDNA clones of "brain-specific" mRNAs using a Northern blot screening procedure (Milner and Sutcliffe, 1983; Sutcliffe et al., 1983b). More recently, we have used subtractive hybridization procedures to select cDNA clones of mRNAs that are differentially expressed during development (Miller et al., 1987) or are expressed in one brain region but not another (Travis et al., 1987; Travis and Sutcliffe, 1988). This chapter will discuss the methodologies available for the analysis of such clones and some of the thinking that underlies their application. Because the generation of cDNA libraries and selection of clones by subtractive hybridization is described extensively in other chapters, we will not discuss these aspects here but assume that the starting point is a clone selected from a cDNA library.

2. Patterns of mRNA Expression

An essential first step in characterizing any cDNA clone is determination of the properties of the corresponding mRNA, particularly its size, abundance, tissue distribution, and temporal expression during development. In subtractive hybridization and other selective cloning procedures, these data are necessary to show that the corresponding mRNA has the desired properties (e.g. tissue or regional specificity of expression) as well as to decide which of several clones are most interesting for further study. The expression pattern of the mRNA may also give important clues as to the type of protein that it encodes: Where is it expressed in the body? Where is it expressed in the brain? When and where is the mRNA first expressed during development?

These data are most easily obtained by RNA (Northern) blotting procedures but a much more detailed picture of the cellular expression of the mRNA can be obtained by *in situ* hybridization. The technical details for both of these procedures are given in the chapters by Morrison and Wilson, respectively; we will focus here on their use in characterizing a new clone and providing clues about its encoded protein.

2.1. Northern Blotting

The development of techniques enabling transfer of RNA to nitrocellulose (Thomas, 1980) was a dramatic improvement over previous procedures using derivatized papers and made Northern blotting accessible to many nonmolecular biologists. In our labs this is often the first "molecular" technique that we introduce to someone without molecular training. Interpretable results are absolutely dependent on the preparation of undegraded extracts of

Analysis of Brain-Specific Gene Products 81

RNA as well as on attention to the details of blotting and hybridization. The wealth of information about mRNA that can be gained from Northern blots gives this technique a considerable advantage over procedures such as dot blotting, which can only indicate presence of the mRNA and its approximate abundance.

Several important pieces of information about an mRNA can be learned from a Northern blot. The position of the hybridizing band on the blot, relative to size standards, provides the approximate size of the mRNA. This can be compared to the size of the cDNA insert to determine what fraction of the mRNA is represented in the clone, and, hence, whether it will be necessary to screen for additional clones. Northern blots may also reveal several mRNA species that hybridize to a particular cDNA probe, indicating the existence of several transcripts derived from a single gene or a cross-hybridizing family of mRNAs derived from several genes. Such complexity would not be detectable with dot blots.

The intensity of the hybridization signal is an indication of the abundance or concentration of the mRNA in the tissue. It is possible to make reasonable estimates of the abundance of a particular mRNA, if the blots are performed carefully and consistently, with attention to probe concentration and specific activity, hybridization time and temperature, and autoradiographic exposure time, and compared to similar blots using probes to mRNAs of known abundance. More accurate estimates can be made by including on the blot known amounts of the cDNA clone or cRNA made by transcription of the cloned insert. Northern blots can be used to detect mRNAs with concentrations lower than 0.001% (roughly 2–5 molecules/cell). Some brain mRNAs, particularly those that are expressed in only a few cell types, may have an overall abundance in total brain of considerably less than this and may not be detectable on Northern blots of whole brain RNA samples. To detect these mRNAs, it may necessary to use RNase or S1 nuclease protection assays, which measure the ability of the mRNA to hybridize to and protect a radioactively-labeled complementary RNA or DNA strand.

Last, Northern blots can be used to survey the expression of a mRNA in different tissue sources. This survey may also identify a more abundant source of the mRNA than that used for construction of the original cDNA library and should, therefore, be carried out at an early stage in the characterization of a mRNA. A collection of good quality RNA preparations from different sources is, therefore, an extremely valuable resource and might include RNA samples from different tissues (e.g., brain, liver, kidney, pituitary, adrenal gland), from different regions of brain, from brains or other tissues of animals at different pre- and postnatal developmental stages, from animals after drug administration or other experimental manipulations, and from cultured cell lines. The quality and quantity of the RNA samples must be evaluated by analysis with a control cDNA clone that hybridizes to a mRNA present in each sample. Because Northern blots can be run with either total cellular RNA or poly $(A)^+$ samples, these preparations can be stored and used in either form. Also, the blots themselves can be reused several times and can be stored, preferably at low temperature in sealed bags containing desiccant.

2.2. In Situ *Hybridization*

The expression patterns of mRNAs can also be determined by *in situ* hybridization. Although this technique is less easily quantifiable than Northern blotting, it also has some advantages, particularly for determining the cellular sites of mRNA expression. *In situ* hybridization experiments can be used as part of a screening process to determine rapidly the anatomical distribution of cloned mRNAs, in order to choose among several of interest. To facilitate this procedure, cDNA libraries can be prepared in plasmid or bacteriophage vectors (e.g., "Gemini" vectors from Promega or Lambda ZAP and pBS vectors from Stratagene) that contain bacterial RNA polymerase promotors positioned so that cRNA probes ("riboprobes") for *in situ* hybridization analysis can be generated directly without further subcloning.

Most *in situ* hybridization procedures use radioactively labeled probes, although there are methods for nonradioactive detection based on incorporation of biotin-derivatized nucleotides. The hybridization of radioactive probes can be detected either by placing the section against X-ray film to obtain a gross image of the hybridization pattern or by coating the section with photographic emulsion to give cellular level resolution. Both approaches are useful and with strong β emitting isotopes, such as ^{32}P or ^{35}S, it is feasible to do both on the same section sequentially, giving an overview of the pattern of expression and possibly identifying the cell types responsible. Initial screens of a clone of a brain mRNA might, therefore, include sagittal sections of adult brain or a series of spaced coronal sections, to give the regional pattern of expression in the brain, and sagittal sections of whole embryos, to provide the pattern of expression during development.

The expression pattern for a particular mRNA, obtained by a combination of Northern blotting and *in situ* hybridization, may indicate the kind of functions that the encoded protein might mediate. A correspondence between the distribution of the mRNA and the distribution of a known neurochemical marker, neurotransmitter system or cell type may immediately suggest potential functions. For example, localization to white matter areas, particularly during late postnatal development, would indicate involvement with myelination; expression in the cell bodies of single source distributed systems, such as the norepinephrine innervation arising from the locus ceruleus, might indicate association with the corresponding neurotransmitter system. Ideally, one would like to be able to compare a novel distribution pattern to all known patterns: the problem of accessing this body of information is discussed in Section 6.3.

3. Interpretation of Nucleotide and Amino-Acid Sequences

A prime goal of many cloning projects is to determine the particular properties of a protein of interest and to relate these to the possible functions of the molecule. The key to the structure of a protein is contained within its amino acid sequence, which is generally thought to determine its three-dimensional folding pattern. Unfortunately, we still have only a primitive understanding of the rules that govern the relationships between the shape of a protein and its linear primary sequence. But this situation can be expected to improve as more protein sequences are determined and related to other sequences and to three dimensional structures determined by X-ray crystallography. The accumulation of new protein sequence information has been in large part the result of the determination of nucleotide sequences and can be expected to increase exponentially as automated nucleotide sequence determination (Smith et al., 1986; Prober et al., 1987) becomes a routine laboratory service. Carefully solved nucleotide sequences can be highly reliable, because the techniques for their generation are relatively free of the systematic artifacts that can plague protein sequence determination.

We cannot overemphasize the importance of complete sequence information. The first cDNA clones isolated will usually be partial copies of the mRNA. These can be used for preliminary characterization of the expression patterns of the mRNA by RNA blotting and *in situ* hybridization assays, as described above, and this information is often necessary to decide whether a clone is worthy of further study. Once a decision has been made to study a particular clone in detail, however, the first priority should be the isolation and characterization of additional clones to complete the structure of the mRNA. Ideally, the 5' terminus of the mRNA should also be mapped by primer extension experiments in order to determine if selected cDNA clones are full length copies of the mRNA. In addition, a single gene can generate multiple mRNA products that differ from one another at their 5' ends because of alternative initiation sites for transcription, at their 3' ends because of polyadenylation at different sites, or at internal regions because of alternative RNA splicing. Some of these differences may be revealed by broad or multiple bands on Northern blots, but others may only surface by careful inspection of cDNA clones for differences in the gel migration patterns of restriction endonuclease generated fragments of the cloned cDNAs.

3.1. Definition of Open Reading Frames

A eukaryotic mRNA contains an internal region, known as the *coding region* or *open reading frame (ORF)*, which is translated into protein. The ORF is flanked by *noncoding* or *untranslated regions*, and is usually located closer to the 5' end of the mRNA (Fig. 1). For example, for 39 characterized brain mRNAs with an average length of 3243 nucleotides, the average 5' untranslated region was 139 nucleotides (range 7–443), the average ORF was 1385 (range 222–6027), and the average 3' untranslated region was 1113 nucleotides (range 57–2328) (Sutcliffe, 1988). One of the two strands of a double-stranded cDNA, known as the *sense* or *plus* strand, will correspond in sequence to the mRNA from which it was derived and will provide the protein sequence. (The cDNA strand complementary to the mRNA sequence is usually called the *antisense* or *minus* strand.) For a cDNA clone that includes the 3' terminus of the mRNA, the sense strand will usually be indicated by the poly(A) tail. In other cases, identification of the sense strand may need to be made experimentally, for example, on Northern blots by

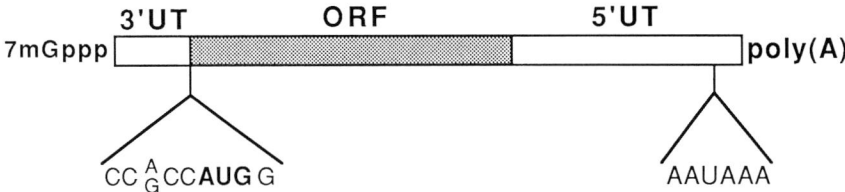

Fig. 1. Structure of a typical eukaryotic mRNA. The open reading frame (ORF) is flanked by 5' and 3' untranslated regions (UT). A 7-methylguanosine residue is attached to the 5' terminus by a triphosphate linkage and the RNA terminates at the 3' end with a poly (A) tail. A consensus sequence around the initiation codon and the polyadenylation signal are also shown.

using singled stranded-probes directed against each strand. Such probes are readily available if the cDNA insert has been cloned into M13 or similar vectors.

The sequence of the protein encoded by a cDNA is derived by translating the nucleotide sequence of the cDNA using the genetic code. Translation of nucleotide into amino-acid sequences is a common feature of computer sequence manipulation programs, which are now available for most personal and mainframe computers, although the same result can be obtained more laboriously by hand. Because the genetic code is based on nucleotide triplets, three different reading phasings are possible on each DNA strand: one of these will contain the ORF that encodes the protein. The ORF begins with a methionine codon (AUG), and is usually the longest uninterrupted frame from the three phasings. Generally, identification of the ORF is unambiguous if the cDNA sequence is full-length. Occasionally, there may be some ambiguity about the start of the frame when additional AUG triplets are present near the 5' end of the ORF: in these cases, it may be necessary to verify the correct start site by determining the amino terminal sequence of the protein product.

3.2. Consensus Sequences

Several recognizable "consensus" sequences that act as sites of regulation or modification have been identified by comparative analysis of the sequences of mRNAs and their encoded proteins. In some cases, the definition of a consensus sequence has been refined by mutagenesis and gene construction experiments. For mRNAs, consensus sequences are useful for defining regions of the molecule; for proteins, they generally indicate sites of posttranslational modification.

3.2.1. Consensus Sequences in mRNAs

Apart from the coding region of an mRNA, where the significance of the information content is obvious, relatively few consensus sequences have been defined for eukaryotic mRNAs. Each eukaryotic mRNA has a "cap" at its 5' end, a 7-methylguanosine residue attached by a triphosphate linkage. The cap structure is required for initiation of protein synthesis on ribosomes. Translation begins, usually but not always, at the first in frame AUG codon. The consensus sequence, $CC^A/_GCCAUGG$, has been proposed for the region surrounding the initiation codon (Kozak, 1984), with the sequence ACCAUGG giving optimal initiation on eukaryotic ribosomes (Kozak, 1986). It has been proposed that this sequence hybridizes to its complement at the 3' terminus of eukaryotic 18S ribosomal RNA (Sargan et al., 1982; Kozak, 1986), in a manner analogous to the "Shine-Delgano" sequence of prokaryotic mRNAs.

The 3' untranslated region will usually contain a polyadenylation signal sequence, AAUAAA, some 10–30 residues upstream of the poly(A) tail (Birnstiel et al., 1985). Minor variations in this sequence are common. A single gene may contain multiple polyadenylation signals, generating a family of mRNAs that differ in the lengths of their 3' untranslated region, as has been shown to occur for more than 25% of brain mRNAs (Sutcliffe, 1988). In these cases, the longer mRNAs will contain the unutilized polyadenylation signals for the shorter mRNAs. Other sequences, adjacent to the polyadenylation signals but as yet poorly defined, may also influence the choice of polyadenylation site. The function of the 3' untranslated region in general is also unclear, although AU rich sequences, such as AUUUA, may determine mRNA stability (Shaw and Kamen, 1986).

3.2.2. Consensus Sequences in Proteins: Sites for Modification

This discussion will be concerned with those sequences that are common to many families of proteins and define sites of posttranslational modification. The presence of these sequences can indicate the kind of protein: membrane-bound, secreted, proteolytically processed, and so on. Sequences that may define particular three dimensional structures are discussed in Section 4.2.4.

3.2.2.1. SIGNAL PEPTIDE SEQUENCES. Proteins that are destined to be secreted or to be anchored in a membrane often contain a sequence of 15–25 amino acids, known as a signal peptide, at the amino terminus of the ORF. This sequence facilitates the translocation of the nascent polypeptide chain into the lumen of the endoplasmic reticulum and, hence, directs the protein to the Golgi complex and the secretory pathway (Blobel, 1980). The signal peptide is removed from the nascent polypeptide by specific endopeptidases as soon as it enters the lumen of the endoplasmic reticulum. The sequence of the signal peptide is generally hydrophobic, containing stretches of leucine and valine residues but without any defined consensus (von Heijne, 1985). There are usually one or two positively charged residues at the amino terminus, and cleavage often occurs on the carboxyl side of an amino acid with a small sidechain, such as alanine, glycine or serine (von Heijne, 1984). As discussed below, some membrane-associated proteins may use internal signal sequences that are not proteolytically removed after translocation.

3.2.2.2. TRANSMEMBRANE DOMAINS. An integral membrane protein will have one or more regions that cross or are otherwise associated with the lipid bilayer of a cellular membrane. Like the signal peptide, these regions are very hydrophobic and usually lack any charged residues. A length of 21 amino acids is considered sufficient to cross the average biological membrane (Eisenberg, 1984). Transmembrane domains can be detected by scanning the protein sequence: several computer algorithms will calculate the average "hydrophilicity" for a moving window of defined sequence length and plot the result as the hydrophilicity profile of the protein (Hopp and Woods, 1981). For example, the profile for the brain protein 1B236/MAG, which is assumed to have a single transmembrane domain, shows a predominant, hydrophobic peak close to the carboxyl terminus (Fig. 2): the corresponding sequence consists of 21 uncharged residues, with a preponderance of aliphatic side chains (leucine, isoleucine, valine). Several positively charged residues are often found at the cytoplasmic boundary of many transmembrane domains, as is shown here for 1B236/MAG.

Some membrane associated proteins with an externally oriented amino terminus do not possess a transmembrane domain

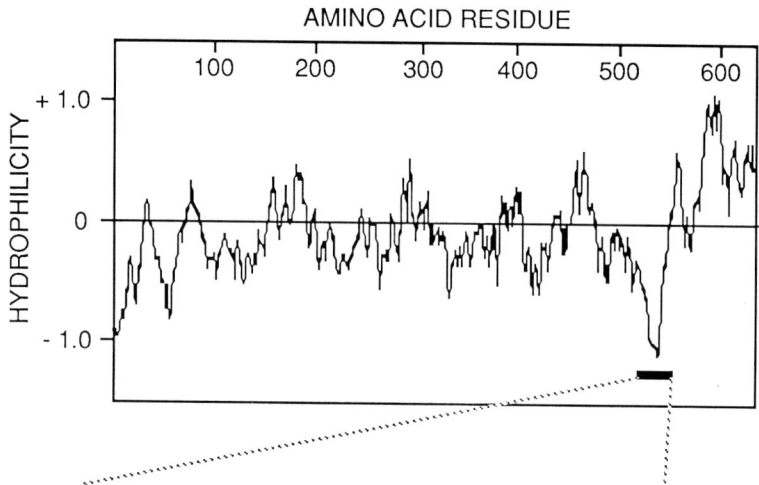

TrpAlaLysIleGlyProValGlyAlaValValAlaPheAlaIleLeuIleAlaIleValCysTyrIleThrGlnThrArgArg

Fig. 2. Hydrophilicity plot for the protein 1B236/MAG. The sequence of the developmentally early form of 1B236/MAG (Lai et al., 1987a) was analyzed by the procedure of Hopp and Woods (1981), computing the average hydrophilicity in a moving window of 21 residues (analysis performed on an Apple Macintosh using the program DNA Inspector). The sequence corresponding to the peak of greatest hydrophobicity is shown below, with the putative transmembrane domain underlined.

but are conjugated directly to membrane lipids by a phospholipid linkage. In the studied examples, which include a form of the cell adhesion molecule N-CAM, a precursor contains a hydrophobic carboxyl terminal sequence, similar to a transmembrane domain but without any following charged residues. This sequence may be cleaved from the nascent polypeptide as the new carboxyl terminus is conjugated to phospholipid.

Membrane proteins have been classified into three types based on their orientation and topology in the membrane (Fig. 3) (Garoff, 1985). Group A proteins contain a single transmembrane domain and are oriented with their amino termini on the outer face and their carboxyl termini on the inner or cytoplasmic face: neural examples of this class include N-CAM (Cunningham et al., 1987), and the myelin proteins P_o (Lemke and Axel, 1985) and 1B236/MAG (Lai et al., 1987a). Such proteins usually use an amino terminal signal peptide sequence to direct the molecule into the

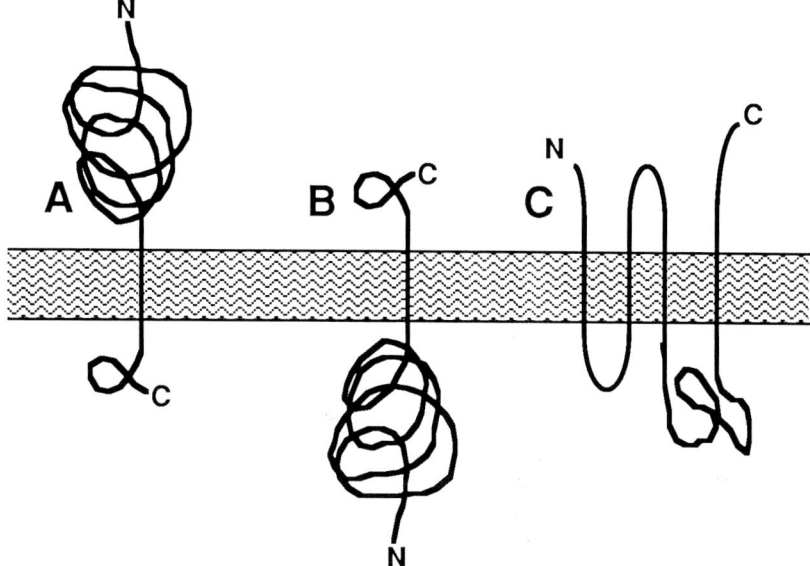

Fig. 3. Classes of membrane-associated proteins. These are divided into three groups according to their orientation and topology (Garoff, 1985). Group A: single transmembrane domain, amino terminus outside; Group B: single transmembrane domain, amino terminus inside; Group C: polytopic proteins with multiple transmembrane domains, the termini may be on either side of the membrane.

lumen of the endoplasmic reticulum; the transmembrane domain then anchors the protein correctly in the membrane.

Group B proteins also contain a single transmembrane domain but have the opposite orientation, with the amino terminus on the cytoplasmic face and the carboxyl terminus projecting externally. Examples of this type include the liver asialoglycoprotein receptor and the transferrin receptor. Group B proteins generally do not possess cleaved amino terminal signal peptides: in the studied examples, the single transmembrane domain has been shown to have a dual function as a internal signal sequence as well as a membrane anchor.

Group C proteins are polytopic, crossing the membrane several times with multiple transmembrane domains. Examples include opsin (Friedlander and Blobel, 1985), the β-adrenergic and muscarinic receptors, and, probably, myelin proteolipid protein. These proteins may or may not utilize signal peptide sequences;

their correct orientation in the membrane is defined by multiple internal signal sequences and so-called "stop-anchor" sequences adjacent to the transmembrane domains, as has been shown for opsin (Friedlander and Blobel, 1985). The amino termini of group C proteins can be located on either membrane face.

The identification of putative transmembrane domains may suggest models for the topology of the protein in the membrane. These hypothetical models can be tested experimentally by translation of the corresponding mRNA in vitro in the presence of microsomal membranes (see chapter by Colman). In the many examples studied, the nascent polypeptide is correctly inserted in the membrane, and its orientation can be investigated by sensitivity of exposed regions to proteases, as well as by mutagenesis and gene reconstruction experiments (Garoff, 1985).

3.2.2.3. GLYCOSYLATION SITES. Carbohydrate sidechains are added to asparagine residues (N-linkage) within the consensus sequence Asn - X - Ser/Thr, where X is any amino acid (Neuberger et al., 1972). Not all such sequences within a protein are necessarily glycosylated, and other secondary structural features may influence which sites are used. In addition, glycosylation of proteins occurs within the lumen of the endoplasmic reticulum or Golgi apparatus; therefore, a polypeptide must also possess signal peptide sequences (amino terminal or internal) to target the nascent chain appropriately, and glycosylation will only occur on those regions of the chain that enter the endoplasmic reticulum. Presence of these sequences definitely warrants tests for glycosylation of the encoded protein. For example, there are nine potential sites for N-glycosylation within the amino acid sequence of the protein 1B236/MAG (Lai et al., 1987a). Eight of these are in the putative extracellular part of the molecule, and experiments show that the native molecule contains 25–30,000 daltons of carbohydrate (Sutcliffe et al., 1983a), suggesting that all or most of these sites are used. A ninth site is in the putative cytoplasmic tail, which is unlikely to enter the endoplasmic reticulum, and is, therefore, probably not glycosylated. No consensus sites have been defined for O-linked glycosylation to serine or threonine residues.

3.2.2.4. PHOSPHORYLATION. Proteins may be phosphorylated on the hydroxyl groups of serine, threonine, or tyrosine residues. One group of enzymes, which includes cAMP-dependent and calcium/calmodulin-dependent protein kinases and protein kinase C, acts predominantly on serines with some phosphorylation on threonines (Nestler and Greengard, 1984). The phosphorylation site for these enzymes usually follows several basic residues. For cAMP-dependent kinase, the consensus sequence around a phosphorylation site has the general form, . . .Arg - Arg - X - Ser/Thr . . ., where X can be any amino acid but is most commonly an uncharged amino acid. For calcium/calmodulin-dependent protein kinase II, the sequence is very similar, with the form, . . .Arg - X - Y - Ser/Thr . . ., where X and Y can be any amino acid. Both types of phosphorylation site are represented, for example, in the neural protein synapsin I (Czernik et al., 1987). It should be emphasized, however, that very few phosphorylation sites have been characterized. Furthermore, there are a large number of protein kinases that are currently being characterized. Many of these have quite precise substrate specificities, indicating that other structural features in the protein may determine which particular sites are targets for phosphorylation by each particular enzyme.

A second group of enzymes—tyrosine kinases—catalyze the phosphorylation of tyrosine residues (Hunter and Cooper, 1985), including a tyrosine within their own catalytic domain (autophosphorylation). Some sites, including the major autophosphorylation site of pp60^{v-src}, have the general structure, . . . Arg - X - X - X - Glu - X - Glu - Tyr . . ., but, again, few such sites have been defined and the consensus is not strong.

3.2.2.5. PROTEOLYTIC CLEAVAGE SITES   Neuropeptides and other biologically active peptides are generated by proteolysis of larger precursor molecules (Loh et al., 1984); often several active peptides may be derived from a single precursor (Douglass et al., 1984). In the majority of cases, proteolytic cleavage occurs at pairs of basic amino acids, most commonly at Lys–Arg but also at Lys–Lys, Arg–Arg, and Arg–Lys sequences. Less frequent cleavages occur at single basic residues or at other sites. The existence of pairs of basic residues within a protein sequence does not necessarily imply that the protein is proteolytically cleaved to generate active peptides: for example, pairs of basic residues occur frequently in proteins such as histones or myelin basic protein, which are not known to undergo proteolytic processing.

Some peptides also carry amide groups (–$CONH_2$) on their carboxyl ends: the amine (NH_2) group of the amide is derived from the amine group of a glycine residue on the amino terminal side of the cleavage site (Bradbury et al., 1982). Thus, the sequence,. . . Pro-Arg-Gly-Gly-Lys-Arg-Ala . . ., in the precursor for vasopressin generates the carboxyl terminal sequence of mature vasopressin, . . .Pro-Arg-Gly-amide (Land et al., 1982).

3.2.2.6. OTHER MODIFICATIONS. Proteins can also be modified by acetylation at the amino terminus or on lysine residues, by formation of an amino terminal pyroglutamate, by sulfation of tyrosine residues (usually those following acidic residues; Lee and Huttner, 1983), or by acylation with long chain fatty acids, but no well-defined consensus sequences have as yet been determined for these reactions.

4. Sequence Comparisons

It is obviously important to determine, at an early stage, whether a newly defined protein sequence corresponds to one that has been previously characterized. This can be accomplished by searching the available computer databases of known sequences. The search methods have considerable flexibility and can generally identify similar as well as identical sequences. Thus, it is possible to determine whether a protein is related to any others, such as the same protein from a different species or enzymes with similar catalytic functions. Other programs enable two particular sequences to be compared and provide a measure of the statistical significance of any sequence similarities. In this fashion, families of proteins with common structural features can be established. Any of these analyses may identify a protein or suggest possible functions.

4.1. Searching the Databases

Computer databases have been established for both nucleic acid and protein sequences. It is possible to use either type, but we do not recommend searching nucleic acid databases for similarities with coding region (i.e., protein) sequences. Nucleic acid searches tend to be slower, because of the larger size of the database, and the results are often difficult to interpret because of the degeneracy of the genetic code. In the US, a protein database (NBRF-PIR) is maintained by the National Biomedical Research Foundation (Barker et al., 1987); in Europe, the protein database PseqIP (Claverie and Bricault, 1986) is a nonredundant compilation of four independent databases, including NBRF-PIR. These databases, as well as programs for their analysis, can be obtained from their guardians or from commercial sources (e.g., Intelligenetics, DNAStar). Both contain the sequences of several thousand different proteins, with greater than one million total amino-acid residues.

Several different computer algorithms have been developed for database searching. The SEARCH program from NBRF (Dayhoff et al., 1983) for example, compares the test sequence against all possible alignments in the database, each alignment is scored by summing defined values for each pair of aligned residues, and the highest scoring alignments are listed in an output file. The program FASTP (Lipman and Pearson, 1985) first looks for local groups of identities between the test sequence and each sequence in the database and then optimizes the alignment. Searches for distant relationships are facilitated by using a scoring matrix (PAM250) derived from comparisons of changes in the sequences of known related proteins, reflecting the probabilities of amino-acid replacement during evolution and giving higher values to matches of the most conserved amino acids, such as tryptophan or cysteine (Dayhoff et al., 1983). A different type of algorithm, devised by Brenner (1987), searches for defined consensus sequences using sequence "templates" in which each element must be matched by a given amino acid, one of several amino acids, any amino acid, or any except certain amino acids.

4.2. Assessment of "Homology"

In the most fortunate (or sometimes unfortunate!) case, a database search may find an identical or very similar sequence, giving an immediate identity to the sequence used for the search. More frequently, however, database searches identify a bewildering array of potentially related sequences. Because of the large size of the databases, many of these result from chance alignment of residues. How can the significance of these similarities be assessed?

4.2.1. Evolutionary Changes in Protein Sequences

First, let us consider how evolutionary change can generate differences between protein sequences. As species diverge from common ancestors during evolution, so also do the sequences of individual genes and their protein products. In the simplest case, the degree of difference between the sequences of the same gene or protein in any two species is related to the evolutionary distance between them. Particular genes diverge at markedly different rates, however, depending on the structural and functional constraints on their protein products. For example, the amino-acid sequences of myelin proteolipid protein sequences vary by only 1–2% among mammalian species, the sequences of myelin basic proteins show roughly 10% variation, and those for the secretogranin, chromogranin B, show 35% variation.

More distant relationships may occur when proteins with different functions evolve from a common ancestor. For example, the enzymes tyrosine hydroxylase and phenylalanine hydroxylase catalyze reactions of structurally similar substrates. Rat tyrosine hydroxylase (Grima et al., 1985) and human phenylalanine hydroxylase (Kwok et al., 1985) have identical residues at 53% of their positions. The similarities in sequence, as well as in the sizes and functions of these proteins, strongly suggests that they arose from a common precursor. Analysis of these more distant similarities leads to the definition of protein families and superfamilies that share sequences and common structural features, as described below.

A word here about the much misused term "homology." As recommended by Reeck and colleagues (1987), this term should be used to mean "possessing a common evolutionary origin" and not be used as a synonym for sequence similarity. Postulating homology between two proteins requires several lines of evidence, including a significant level of sequence identity. Tyrosine and phenylalanine hydroxylases, for example, can properly be called homologous proteins, based on the evidence described above, but it is incorrect to refer to their similarity as 53% homology: 53% identity is the correct description.

4.2.2. Analysis of Sequence Similarities

How, then, should sequence similarities be assessed? A first consideration is the length of the similarity between sequences. Does similarity occur only in small regions of the sequences or does it extend throughout their lengths? Are the identical residues clustered and do they occur in the same relative parts of the two proteins? Other clues can be obtained from the nature of the similar sequences: are these from distant or closely related species? Is the test protein likely to have similar functions analogous to those of the similar proteins?

Finally, the significance of any alignments can be tested statistically. The most common measure of significance, often called the *alignment score* or *Z-value*, compares the score for the best alignment between two sequences with scores for alignments with one of the sequences randomized. The program ALIGN (Dayhoff et al., 1983), for example, determines the score for the best alignment between two sequences, using the PAM250 matrix. It then randomizes one of the two sequences and again determines the score for the best alignment; this is repeated, usually one hundred times, to provide a mean score and standard deviation for the alignments of the randomized sequences. The alignment score, or Z-value, is then calculated as follows:

$$Z = \frac{\text{test score - mean of random score}}{\text{standard deviation of random score}} \quad (1)$$

Higher Z-values indicate a higher probability that the alignment could not have occurred by chance: usually, values greater than 3–4 are considered significant. Z-values of 3, 6, and 9, for example, correspond to probabilities of 0.135×10^{-2}, 0.987×10^{-9}, and 0.113×10^{-18}, respectively, that the alignment did not occur by chance (Dayhoff et al., 1985). For example, the Z-value for an alignment between human phenylalanine hydroxylase and rat tyrosine hydroxylase is 52.8, a highly significant similarity and excellent evidence that the proteins are homologous.

4.2.3. Dot Matrix Analysis

Similarities in protein or nucleic acid sequences can also be displayed by the dot matrix method. In the program DIAGON (Staden, 1982), a moving window of given length from one sequence is compared to all positions in the second sequence; positions at which there are similarities are plotted as dots on a two dimensional matrix. Proteins that are closely related in sequence give a strong pattern of dots along the diagonal of the matrix. This method is also particularly good in revealing repeated sequence patterns within a protein. When a sequence is compared with itself, repeated motifs will be displayed as lines of dots parallel to the diagonal of identity.

4.2.4. Families and Superfamilies

The analysis of similarities among protein sequences has revealed groups of proteins that appear to be related. Dayhoff has defined several categories, depending on the degree of similarity (Dayhoff et al., 1983): *subfamilies* contain proteins that are less than 20% different in sequence; *families* are less than 50% different in sequence; and *superfamilies* contain proteins with similarities that have a probability of less than 10^{-6} of occurring by chance (approximately equivalent to a Z-value of 4.5–5). By this classification, for example, the enzymes tyrosine hydroxylase and phenylalanine hydroxylase, which differ among mammalian species by approximately 50%, form a protein family. Within this family are subfamilies containing representatives of each of the two enzymes, which each differ among mammals by 5–10%.

The similarities in the primary sequences of members of families and superfamilies imply that these proteins also share common features of three-dimensional structure. For example, a notable protein superfamily is the immunoglobulin superfamily (Williams, 1987), which has several members expressed in the nervous system, including 1B236/MAG, the neural cell adhesion molecule N-CAM, and the peripheral myelin protein P_o (Lai et al., 1987b). Each of these proteins contains one or more immunoglobulin-like domains, a sequence of 90–110 amino acids, which in immunoglobulin domains is known to be folded into two opposed β-pleated sheets. Immunoglobulin domains are often involved in domain–domain interactions. Thus, the shared immunoglobulin domain-like sequences suggest similar three-dimensional structures will be formed in these molecules and that these structural motifs may underlie the potential functions of MAG, N-CAM, and P_o in cell–cell interactions and adhesion. Other shared sequence motifs that underlie functions common to a protein superfamily include the so-called "E–F hand" (Kretsinger, 1976), which forms the calcium binding site of the calmodulin superfamily, and "zinc fingers" (Klug and Rhodes, 1987), a sequence motif common to nucleic acid-binding proteins such as transcription factors.

In this way the classification of proteins and the definition of common structural motifs may reveal potential functions for proteins whose newly defined sequences qualify them as members of one of these families. In some cases the analysis may reveal quite unexpected relationships across large evolutionary distances. Perhaps the most dramatic recent example was the detection of a sequence relationship between mammalian epidermal growth factor and the products of the Drosophila gene *Notch* (Wharton et al., 1985) and the nematode gene *lin-12* (Greenwald, 1985), both implicated in neural development. More superfamily relationships can be expected as the sequences of a larger portion of proteins are determined.

5. Preparation of Antibodies to Encoded Proteins

Although much information about a protein can be deduced from its amino acid sequence, it is obviously necessary to be able to study the protein itself. Antibodies are the best reagents for protein characterization, offering specific detection of the protein in tissue extracts or on sections, quantitation by radioimmunoassay, a means to isolate the molecule by immunoaffinity chromatography, and the possibility of perturbing the function of the protein. Antiprotein antibodies are normally prepared by immunization of animals with purified preparations of the protein. However, for proteins that are defined only by nucleotide sequences, there are alternative approaches for producing antigens that do not require purification of the protein from its natural source: the use of chemically synthesized peptides corresponding to regions of the protein sequence or the biological expression of the protein in various vector–host systems from its cloned nucleotide sequence.

The alternative methods have their strengths and weaknesses. Antipeptide antibodies are very specific, react with defined sites that are chosen by the investigator and, as discussed in Section 6, offer the most reliable method of proving the existence of a hypothetical protein defined by a cDNA sequence. This method can also be routinely applied to any type of protein. However, there is no guarantee that antibodies that are reactive with a particular synthetic peptide will also react with the native protein. This can be a major disadvantage when dealing with proteins translated from cDNA sequences with no knowledge of the properties, abundance, or distribution of the protein in vivo. On the other hand, one can have confidence that the protein generated by an expression system does indeed correspond to the ORF defined by the nucleotide sequence of the cDNA. It is often difficult, however, to obtain efficient expression and this may require custom DNA engineering for each protein. In fact, the best strategy may be to use both approaches. In particular, the expressed protein will provide a positive control to test for reactivity of antipeptide antibodies to the intact molecule.

5.1. Antibodies to Synthetic Peptides

When an animal is immunized with a native protein, antibodies are usually generated against several distinct antigenic determinants or *epitopes* on the molecule. The structure of each epitope is defined by the three-dimensional conformation of a small number of amino-acid residues that are adjacent in space. Particular epitopes may be *continuous,* consisting of contiguous residues on the polypeptide chain, or *noncontinuous,* consisting of amino acids from different regions of the molecule that have been brought together by the three-dimensional folding of the native protein. We can mimic natural continuous epitopes by immunizing not with the whole protein but with short synthetic peptide fragments that correspond in sequence to a region of the protein. A large number of studies have now demonstrated that the resulting antipeptide antibodies bind to the corresponding region of the native protein as well as to the peptide used for immunization. The advantages of this approach are that it is not necessary to purify the protein but only to have knowledge of the protein sequence, and the sites on the protein to which the resultant antibodies react is precisely defined or "predetermined" (Sutcliffe et al., 1983c). Thus, such antipeptide antibodies can serve as chemical probes for the detection of a series of particular known regions of a protein, even if the protein itself is only known putatively as an amino acid sequence translated from a cDNA clone.

5.1.1. Selection of Peptides

Obviously, the key to success in this procedure is in the selection of peptides that correspond to good epitopes in the native protein. The best results have been obtained with peptides corresponding to generally hydrophilic regions of the putative protein (Sutcliffe et al., 1983c); these are most likely to be located on the surface of the protein and thus accessible to antibody. Hydrophilic peptides are also soluble in aqueous solvents and easier to handle. We generally choose peptides of 10–15 residues that contain several (but not exclusively) charged (Lys, Arg, Asp, Glu, His) or small sidechain (Ser, Gly, Ala, Thr) amino acids, taking care to avoid potential sites of glycosylation or other modifications. Peptides containing one or more proline residues appear to be particularly effective: these may have a "hairpin" conformation that is similar to the corresponding region in the native protein. It is important to choose at least two nonoverlapping regions of each protein for peptide synthesis.

5.1.2. Preparation of Antipeptide Antibodies

Once appropriate peptides have been selected and synthesized three experimental steps are necessary to obtain good antipeptide antibodies: preparation of peptide-carrier protein conjugates, immunization of animals, and testing for antibody activity. We give here our current laboratory protocols for the preparation of polyclonal antibodies in rabbits; the same peptide-carrier conjugates could be used, however, for the preparation of monoclonal antibodies in mice.

5.1.2.1. CONJUGATION OF PEPTIDES TO CARRIER PROTEINS. In general, peptides are conjugated to a larger carrier protein for immunization. This increases their immunogenicity and reduces their diffusion from the site of immunization, although some larger peptides can elicit antibodies in an unconjugated form. Carrier proteins are generally large, highly immunogenic molecules, often from evolutionarily distant sources. The protein keyhole limpet haemocyanin (KLH), for example, is widely used for this purpose. In early experiments, however, we found that antisera generated with this carrier often contained carrier-specific antibodies that reacted with brain sections but were unrelated to the conjugated peptide. For similar reasons, bovine serum albumin is an inappropriate carrier to use in mammalian studies, particularly for cultured cells. Consequently, we now use thyroglobulin or the plant protein edestin as carriers. The goal of the conjugation reaction is to decorate the carrier protein with peptides but to avoid large aggregates in which most peptides are not exposed to solvent. Synthetic peptides are conjugated to edestin and other carriers in the following manner:

1. Dissolve 500 nmoles peptide (approximately 5 mg) in 0.7 mL 0.25 M phosphate buffer, pH 7.4. Adjust pH if necessary to dissolve peptide, then readjust to pH 7.4.

2. Add 5 mg carrier protein (KLH, edestin, thyroglobulin).
3. Add 4 μL 25% glutaraldehyde. (For some peptides, this amount of glutaraldehyde may result in the formation of a gel. If so, repeat with a diluted glutaraldehyde solution, adding glutaraldehyde until the reaction mixture becomes slightly cloudy.)
4. Incubate 30 min at room temperature, shaking occasionally.
5. Load onto Sephadex G-50 column (0.7 × 30 cm), equilibrated and eluted with 50 mM phosphate buffer, pH 6.0. Collect protein peak. Dialyze overnight against phosphate buffered saline (PBS: 10 mM sodium phosphate, pH 7.4, 150 mM NaCl). If the peptide mixture appears very thick, do not load on a column but remove unconjugated peptide by dialysis against PBS for at least 48 h.
6. Aliquot and store at –20°C.

5.1.2.2. IMMUNIZATION PROCEDURES. Rabbits are typically immunized with the peptide-carrier conjugates according to the following schedule:

Day 0: first injection. Mix 400 μg peptide (conjugated to carrier) in a final volume of 1.5 mL PBS with 1.5 mL complete Freund's adjuvant. Emulsify and transfer to syringe. Inject 1.5 mL per rabbit subcutaneously in flanks. (May also be injected into footpads). Rabbits should be bled prior to first injection to obtain preimmune (control) serum.

Day 14: second injection. Same as first injection but with incomplete Freund's adjuvant (minus mycobacteria).

Day 21: third injection. Mix 400 μg peptide in a final volume of 1.2 mL PBS with 0.8 ml alum [$Al(OH)_3$, 10 mg/mL], made up as described below. Shake well and inject 1 mL per rabbit.

Days 28, 35: Bleed rabbits.

Boosts are the same as the third injection and are routinely given every 4–6 wk, followed by bleeds 7 and 14 d after each injection. If a strong antibody response is obtained, boosts may be eliminated or the dose of peptide decreased.

The alum is prepared as follows:

1. Add 66.6 g aluminum sulfate [$Al_2(SO_4)_3$] to 100 mL distilled water. Warm to dissolve and transfer to 4-L beaker.
2. Slowly add 600 mL 1 M NaOH with stirring. A precipitate will form.
3. Wash precipitate six times with 3 L water. Allow precipitate to settle, pour off supernatant.
4. Mix precipitate well in blender for 5 min. Measure total volume.
5. Determine concentration of precipitate by adding a well-suspended sample of measured volume to a preweighed beaker, evaporating to dryness on a hot plate at low heat and reweighing.
6. Calculate amount water required to bring total concentration of alum suspension to 10 mg/mL. Before adding water to alum, dissolve NaCl in this water to bring final concentration of the alum suspension to 0.85% NaCl. Add NaCl solution to alum, mix well, aliquot, and store.

5.1.2.3. TEST FOR ANTIPEPTIDE ACTIVITY BY ELISA. Sera from the immunized rabbits are first tested for antibody activity against the peptide used for immunization. We routinely use an ELISA for initial tests of antipeptide activity; this assay is performed over three days as follows:

Day 1: Adsorption of Peptide Antigen to Plates.
1. Add 25 µL peptide dissolved in PBS (10 µg/mL) to wells of a 96-well microtiter plate; add PBS alone to control wells. It is most convenient to alternate columns of peptide and control so that dilutions of each serum can be tested in adjacent peptide and control wells. Allow plates to dry overnight at room temperature.

Day 2: Antibody Reaction.
1. Add 50 µL methanol to each well, to fix peptide to plastic. After 5 min, shake off liquid and air dry plate for 10 min.
2. Add 50 µL 3% bovine serum albumin (BSA) dissolved in PBS to each well, to block nonspecific binding

sites. Incubate for 4 h at 37°C in moist chamber. Shake off liquid.
3. Make serial dilutions of antisera as follows: add 45 μL 1% BSA in PBS to top row of wells; add 25 μL to remaining wells. Add 5 μL serum to top row well. Using pipetter set at 25 μL, mix top row well, and transfer 25 μL mixture to second row well. Mix and continue down the column, discarding the 25 μL taken from the last row well. This gives successive 1:2 dilutions of the serum, starting with 1:10 in the first row. (A multitip pipetter is very convenient, if many sera are to be tested.) When dilutions are completed, seal the plate with Parafilm and incubate overnight at room temperature.

Day 3: Detection of Antibody Reaction.
1. Rinse wells ten times with water, making certain that all liquid is removed at each rinse.
2. Add to each well 25 μL of an appropriate dilution of glucose oxidase-coupled goat antibody to rabbit immunoglobulin dissolved in 1% BSA in PBS. (An appropriate dilution of goat antibody will usually be recommended by the supplier.) Incubate 1–1.5 h at 37°C in moist chamber.
3. Rinse wells ten times with water.
4. Add 50 μl developer solution to each well and incubate 45–60 min at room temperature. Developer solution is 0.6 g glucose, 200 μL 1% hydrogen peroxide, 4 mg 2,2'azino-di[3-ethylbenzthiazolinsulfonate(6)] diammonium salt (Boehringer Mannheim, catalog number 756407) dissolved in 28 mL 0.1 M sodium phosphate, pH 6.0.
5. The extent of the color reaction is determined by measuring absorbance at 414 nm using a Titertek Multiscan ELISA plate reader (commonly used in many immunology laboratories). Antipeptide binding is indicated by a more intense color reaction with peptide-coated wells compared to adjacent control wells. A strong antipeptide antiserum will give a positive ELISA result at serum dilutions of 1:1000 or greater.

Active antibodies are then assayed for reactivity against the protein in brain extracts and on brain sections, as described in Section 6.

5.2. Antibodies to Expressed Proteins

An alternative approach is to use a cDNA clone of the mRNA encoding the protein of interest to direct synthesis of that protein in an expression system. Most such systems consist of a *host*, a bacterium or cell line, and a *vector*, which usually contains all or part of a gene that is normally expressed at high levels in the host cells. The cDNA is inserted into the highly expressed gene and the host cells are transfected with the resulting recombinant clone, which is then transcribed and its mRNA translated by the host cell. Depending on the vector and the site of insertion, the foreign protein may be translated as a separate protein or as a fusion protein, covalently linked to all or part of the vector gene product. The desired protein can be recovered and purified from extracts of the cells or from the culture medium. Purification of fusion proteins is facilitated by use of antibodies against the vector gene product. This is almost essential for proteins of unknown identity, for which antibodies or assays may not exist.

Largely because of their importance for the biotechnology industry, a large number of both prokaryotic and eukaryotic expression systems have been developed and are being rapidly improved. Rather than offering detailed protocols for each, we will discuss the major types of expression systems and their strengths and weaknesses.

5.2.1. Bacterial Expression Systems

Among the advantages of bacterial expression systems are the ease of manipulation of host and vector, the ability to grow large amounts of expressing cells, and the absence of any contaminating eukaryotic material in the expressed product. The major disadvantage is that prokaryotic cells do not modify proteins in the same way as eukaryotic cells. For example, proteins are not glycosylated and secreted proteins are not proteolytically processed. This may not be a concern if the product is only destined for antibody production but may be critical if functional activity is required.

Several commonly used cloning vectors can function as expression systems: lambda gt11, for example, generates a fusion protein with β-galactosidase (Young and Davis, 1983). These are widely used for clone selection by protein expression, but they may not be suitable for larger scale synthesis. Other vectors have been designed specifically for high expression and include strong promotor regions and binding sites for bacterial ribosomes (Shatzman and Rosenberg, 1987). Some of these may generate a product that is excreted into the periplasmic space of the bacterial, whereas others require extraction of the bacteria. These often require considerable DNA engineering to obtain the most efficient synthesis of the encoded protein. The genes exploited for expression include β-galactosidase (Ruther and Muller-Hill, 1983), the bacteriophage promotor P_L (Shatzman and Rosenberg, 1987), the "*tac*" promotor (a combination of *trp* and *lac* promotors) (de Boer et al., 1983), and the gene for protein A, which binds immunoglobulin (Nilsson et al., 1985). The major suppliers of molecular biologicals carry extensive lines of bacterial expression vectors. Considerations in the selection of the appropriate system include the size of the protein product, convenient restriction endonuclease sites for insertion of the cDNA to be expressed, and the means at hand to purify the expressed product. Some eukaryotic proteins may be unstable in the bacterial cytoplasm or may even be toxic to the bacteria; such proteins are best expressed by a system that secretes the product into the periplasmic space.

5.2.2. Eukaryotic Expression Systems

The major advantage of eukaryotic expression systems is the high probability that the expressed protein will be processed normally (Cullen, 1987). This is particularly important for secreted or membrane-associated proteins that may require recognition and removal of signal peptide sequences, correct glycosylation, and proteolytic cleavage. Cultured mammalian cells, particularly COS 7, a monkey kidney cell line, have been used extensively for protein expression. Foreign DNA is transfected into these cells by uptake as a calcium phosphate precipitate. In most cases, there is a transient expression of the foreign protein during the first 24–72 h after transfection, and this may be suitable for many screening or functional studies. Constitutive expression, however, requires integration of the transfected DNA into the host cell genome. This is a much less efficient process, occurring in about one in 10^3 cells, and the long term expression of the desired protein may be unstable.

An invertebrate system of great potential is offered by the insect baculovirus, *Autographa californica* nuclear polyhedrosis virus (Lucknow and Summers, 1988). A cultured insect cell line, derived from the fall armyworm *Spodoptera frugiperda*, is used as the host. Vectors have been designed to exploit the gene for the virus polyhedrin protein, which is nonessential for viral replication but is synthesized in very large amounts during the late stages of viral infection and may accumulate to 25% of total cell protein. Results with several mammalian proteins suggest that these are processed correctly: human β-interferon, for example, was glycosylated and secreted in an active form (Lucknow and Summers, 1988).

Proteins have also been expressed in yeast cells with some success: these also appear to be processed correctly (Bitter, 1987). Last, Xenopus oocytes have been used for functional studies of foreign proteins, translated from injected mRNA (Melton, 1987). This system has been used extensively for studies of neurotransmitter receptors, both for selection and identification of cDNA clones of receptors, such as the substance-K receptor (Masu et al., 1987), and for structure-function studies of receptors, using site-directed mutagenesis (*see*, e.g., Mishina et al., 1985). However, this method does not produce sufficient material for antibody production.

6. Characterization of Encoded Proteins

For any protein that is defined solely by a cloned nucleotide sequence an essential first step, once suitable antibodies have been generated, is to prove that this hypothetical protein actually exists in vivo. We have used two experimental approaches: the detection of proteins in tissue extracts using immuno ("Western") blots and detection in brain sections by immunocytochemistry.

6.1. Criteria for Protein Identification

Antipeptide antibodies, with their precisely defined specificities, are particularly suitable for demonstrating the presence of a hypothetical protein. Our criterion for identification of a protein is that the same signal—a band on a Western blot or a pattern of immunocytochemical staining—must be obtained with at least two antibodies directed against independent, nonoverlapping peptides derived from different regions of the amino acid sequence. The rationale is that coincident reactions are much less likely to reflect crossreactions of the antibodies with sequences in other proteins. An additional control is that the reaction of an antipeptide antibody should be blocked in the presence of free peptide, although in our experience, for reasons that are not readily apparent, this does not always occur. As discussed in Section 5, it is also useful to have preparations of the protein generated by an expression system: these can be used as positive controls to demonstrate the activity of antipeptide antibodies and as additional blocking reagents.

These criteria are best illustrated by studies on 1B236/MAG: antipeptide antibodies directed against peptides corresponding to three nonoverlapping regions of the carboxyl terminus of the ORF gave virtually identical staining patterns on brain sections and the reactivity of each antibody was blocked by the appropriate peptide (Sutcliffe et al., 1983b; Bloom et al., 1985). Similarly, all three antibodies detected a 100,000 dalton species on Western blots (Sutcliffe et al., 1983a; Lenoir et al., 1986). The coincidence of these reactions gives very high confidence that the protein being detected does indeed correspond to that defined by the sequence of the cDNA clone.

Once one is confident that the antibodies detect the appropriate protein, the properties of this molecule can be studied in detail. The investigation now enters the realm of standard biochemical and immunocytochemical analyses, most of which are outside the immediate scope of this volume. To complete the description of the analysis of brain-specific genes, however, we give a detailed protocol for Western blotting, the most useful biochemical procedure, and a discussion of strategies to be used for further biochemical and immunocytochemical analyses.

6.2. Biochemical Analysis

6.2.1. Western Blotting

Western or immunoblotting is one of the most useful techniques for rapid characterization of proteins using antibodies. Many variations of this procedure have been documented since it was first described (Towbin et al., 1979). We give here our current laboratory protocol, which is applicable in general cases.

6.2.1.1. SAMPLE PREPARATION. Freshly dissected tissues are rapidly homogenized in 4 vols (w/v) ice-cold 10% sucrose using a Polytron (Brinkmann), mixed with an equal volume of 2X sample buffer (final concentration: 50 mM Tris-HCl, pH 6.8, 2% SDS, 5% 2-mercaptoethanol), heated at 100°C for 5 min, cleared by centrifugation in a microfuge for 5 min, and stored in aliquots at –20°C. Although this method probably solubilizes the majority of cellular proteins, particular proteins may not be extracted efficiently and may require custom extraction procedures. Addition of a cocktail of protease inhibitors (e.g., Trasylol, pepstatin) may be necessary for proteins sensitive to proteases.

6.2.1.2. GEL ELECTROPHORESIS AND BLOTTING. Any derivative of the standard Laemmli (1970) SDS electrophoresis technique is appropriate. An acrylamide concentration of 10% is most convenient for proteins in the 20–100,000 dalton range. We routinely load approximately 50 μg of total protein per lane. Following electrophoresis the gel is removed from the apparatus, trimmed if necessary and incubated for 20–30 min in transfer buffer. The gel is laid on a sheet of Whatman #3 filter paper soaked with transfer buffer and overlaid with a sheet of nitrocellulose that has been presoaked in transfer buffer. It is important to obtain a very close contact between the nitrocellulose and the gel and care should be taken to remove all air bubbles: this can be done by rolling the nitrocellulose with a pipet. The nitrocellulose is overlaid with a second sheet of filter paper and the completed sandwich is placed in the transfer apparatus. We have used two types of apparatus: a BioRad TransBlot apparatus, using 200 mA for 2 hours or a more recently developed apparatus that uses large flat carbon electrodes, allowing the current to be spread more evenly.

6.2.1.3. ANTIBODY BINDING. After transfer the blots are first incubated at room temperature for 1–2 h in a blocking reagent to mask any nonspecific antibody binding sites. We usually use 3% gelatin in TTBS (Tris Buffered Saline: 20 mM Tris-HCl, pH 7.5, 0.15 M NaCl, 0.5% Tween-20), but other protein mixtures can also be used including powdered milk (Johnson et al., 1984). For these incubations, the blots are sealed in a plastic bag ("Seal 'n Save") in a minimal volume of liquid. The blots are then incubated for 2–16 h with antisera in TTBS containing 1% gelatin. The dilution of the antiserum will depend on its titre: strong antibodies can be used at dilutions greater than 1:1000. For initial surveys of activity, dilutions of 1:100 are reasonable. Dilutions of rabbit polyclonal antisera less than 1:100 are likely to give high backgrounds from nonspecific binding or from crossreactions of low affinity antibodies and should be avoided if at all possible. We routinely include nonionic detergents such as Tween-20 or NP-40 in the incubation buffers: this improves detection of some proteins but may also interfere with the reactivity of some antibodies.

6.2.1.4. SIGNAL DEVELOPMENT. The bound antibody is usually detected with an antiimmunoglobulin coupled to an enzyme that will give a colored reaction product. For rabbit antibodies, we use goat antibodies to rabbit IgG coupled to horseradish peroxidase: this gives a purple product in the presence of 4-chloro-1-naphthol and hydrogen peroxide. A variety of signal detection systems are available from commercial suppliers (e.g., BioRad).

1. After incubation with antibodies, transfer blots to a glass dish, rinse briefly with water and wash with three changes of TTBS for 10 min each.
2. Incubate blots with second antibody (usually at concentrations of 1:1000–1:5000) in TTBS containing 1% gelatin in sealed plastic bags for 2 h at room temperature.
3. Transfer blots again to a glass dish, rinse with water, and wash with three changes of TTBS for 10 min each.
4. Incubate blots at room temperature in color development reagent: 90 mg of 4-chloro-1-naphthol (BioRad) are dissolved in 20 mL ice-cold methanol and immediately mixed with 100 mL of TBS containing 60 µL hydrogen peroxide.
5. The colored signal will appear after approximately 5 min; strong signals may appear more rapidly. The blots should not be developed for longer than 15 min. After development, rinse blots for 10 min in running water and dry on filter paper. Potential publication quality blots should be photographed as soon as possible, because the signal fades on storage.

Once a signal has been detected with a particular antipeptide antiserum, the antiserum should next be tested in several dilutions, to determine the working dilution, i.e., the maximum dilution that gives a good signal. That dilution should then be tested for inhibition of the signal in the presence of free peptide (50 µg/mL peptide is usually sufficient). Finally, if several antibodies have been defined for a particular protein, these should be tested for coincidence of their reactions. Samples are run on a gel with a wide slot so that several strips can be cut from a single lane, each of these is blotted and incubated with a different antiserum, and realigned after color development to determine if the bands have identical mobilities. A recent improvement of this method is to clamp a single such blot into a multichambered blotting device (Multiblotter, Immunetics) that provides up to 45 adjacent wells that run the length of the blot; each well can be filled with a different antibody solution. The advantages of this device are that each well holds a small volume, reducing the requirement for antibody, and many lanes appear side by side, allowing easier assessment of coincident reactions. Finally, a single blot can be incubated with two different antisera in the same bag.

6.2.2. Other Antibody-Based Procedures

The antisera generated against peptides or expressed proteins may also be used in many other techniques that employ antibodies to isolate and characterize proteins including, immune precipitation of labeled proteins from cell extracts or in vitro protein translation systems (*see* Colman chapter), radioimmunoassay, and immunoaffinity chromatography. Antipeptide antibodies are particularly useful in dissecting the structure of proteins because they can be generated against particular regions or domains of the molecule.

6.3. Immunocytochemical Analysis

The definition of the anatomical distribution of a novel protein in the nervous system poses potentially enormous problems of data collection and interpretation. In particular, one would like to know if the distribution patterns of the protein correspond to any known neural circuits or neurochemical markers. One solution to this problem is the EMMA system, developed by Bloom and colleagues (Young, Morrison, and Bloom, in preparation), which combines digitized data collection from the microscope with a text and graphics database of all previously described neural circuits. Ideally, the neural database will allow new circuits to be related to old ones, facilitate the discovery of new relationships between circuits, and provide a readily accessible resource of neural information analogous to the protein and nucleic acid databases.

7. Summary: From Structure to Function

We have written this chapter specifically from the perspective of characterizing novel brain-specific proteins that were defined initially by nucleotide sequences derived from cloned mRNAs but the approaches described here are equally applicable to studies of proteins that already have names. In either case, the ultimate goal is to determine the function of the molecule, and to understand how that function is determined by the structure of the protein and by interactions with other components of the cell. Much information about a particular protein can be collected from studies of the expression patterns of its mRNA, from interpretations of the primary sequence of the protein, and from studies of its biochemical properties and neuroanatomical distribution. All of these data can be used to suggest hypotheses for the protein's function, which can be tested experimentally.

Acknowledgments

We acknowledge the National Institutes of Health (NS 20728, NS 21815, NS22347, NS22111) for partial support. This is publication number BCR-5262 from the Research Institute of Scripps Clinic.

References

Barker W. C., Hunt L. C., George D. G., Yeh L. S., Chen H. R., Blomquist M. C., Seibel-Ross E. I., Elzanowski A., Hong M. K., Ferrick D. A., Bair J. K., Chen S. L., and Ledley R. S. (1987) Protein sequence database of the protein identification resource (PIR). National Bio-

medical Research Foundation, Georgetown University Medical Center, 3900 Reservoir Road, N. W., Washington, DC 20007.
Bitter G. A. (1987) Heterologous gene expression in yeast. *Meth. Enzymol.* **152**, 673–684.
Birnstiel M. L., Busslinger M., and Strub K. (1985) Transcription termination and 3' processing: the end is in site! *Cell* **41**, 349–359.
Blobel G. (1980) Intracellular protein topogenesis. *Proc. Natl. Acad. Sci. USA* **77**, 1496–1500.
Bloom F. E., Battenberg E. L. F., Milner R. J., and Sutcliffe J. G. (1985) Immunocytochemical mapping of 1B236: a brain specific neuronal polypeptide deduced from the sequence of its mRNA. *J. Neurosci.* **5**, 1781–1802.
Bradbury A. F., Finnie M. D. A., and Smyth D. G. (1982) Mechanism of C-terminal amide formation by pituitary enzymes. *Nature* **298**, 686–688.
Brenner S. (1987) Phosphotransferase sequence homology. *Nature* **329**, 21.
Claverie J. M. and Bricault L. (1986) PseqIP: a nonredundant and exhaustive protein sequence data bank generated from 4 major existing collections. *Proteins* **1**, 60–65.
Cullen B. R. (1987) Use of eukaryotic expression technology in the functional analysis of cloned genes. *Meth. Enzymol.* **152**, 684–704.
Cunningham B. A., Hemperley J. J., Murray B. A., Prediger E. A., Brackenbury R., and Edelman G. M. (1987) Neural cell adhesion molecule: structure, immunoglobulin-like domains, cell surface modulation, and alternative RNA splicing. *Science* **236**, 799–806.
Czernik A. J., Pang D. T., and Greengard P. (1987) Amino acid sequences surrounding the cAMP-dependent and calcium/calmodulin-dependent phosphorylation sites in rat and bovine synapsin I. *Proc. Natl. Acad. Sci. USA* **84**, 7518–7522.
Dayhoff M. O., Barker W. C., and Hunt L. T. (1983) Establishing homologies in protein sequences. *Meth. Enzymol.* **91**, 524–545.
de Boer H., Comstock L. J., and Vasser M. (1983) The *tac* promotor: a functional hybrid derived from the *trp* and *lac* promotors. *Proc. Natl. Acad. Sci. USA* **80**, 21–25.
Douglass J., Civelli O., and Herbert H. (1984) Polyprotein gene expression: generation of diversity of neuroendocrine peptides. *Ann. Rev. Biochem.* **53**, 665–715.
Eisenberg D. (1984) Three-dimensional structure of membrane and surface proteins. *Ann. Rev. Biochem.* **53**, 595–623.
Friedlander M. and Blobel G. (1985) Bovine opsin has more than one signal sequence. *Nature* **318**, 338–343.

Garoff H. (1985) Using recombinant DNA techniques to study protein targeting in the eukaryotic cell. *Ann. Rev. Cell Biol.* **1**, 403–445.

Greenwald I. (1985) *lin-12*, a nematode homeotic gene, is homologous to a set of mammalian proteins that includes epidermal growth factor. *Cell* **43**, 583–590.

Grima B., Lamouroux A., Blanot F., Faucon Biguet N., and Mallet J. (1985) Complete coding sequence of rat tyrosine hydroxylase. *Proc. Natl. Acad. Sci. USA* **82**, 617–621.

Hopp T. P. and Woods K. R. (1981) Prediction of protein antigenic determinants from amino acid sequences. *Proc. Natl. Acad. Sci. USA* **78**, 3824–3828.

Hunter T. and Cooper J. A. (1985) Protein-tyrosine kinases. *Ann. Rev. Biochem.* **54**, 897–930.

Johnson D. A., Gautsch J. W., Sportsman J. R., and Elder J. H. (1984) Improved technique utilizing nonfat dry milk for analysis of proteins and nucleic acids transferred to nitrocellulose. *Gene Anal. Tech.* **1**, 3–8.

Klug A. and Rhodes D. (1987) 'Zinc fingers': a novel protein motif for nucleic acid recognition. *Trends Biochem.* **12**, 464–469.

Kozak M. (1984) Compilation and analysis of sequences upstream from the translational start site in eukaryotic mRNAs. *Nucl. Acids Res.* **12**, 857–872.

Kozak M. (1986) Point mutations define a sequence flanking the AUG initiator codon that modulates translation by eukaryotic ribosomes. *Cell* **44**, 283–292.

Kretsinger R. H. (1976) Calcium-binding proteins. *Ann. Rev. Biochem.* **45**, 239–266.

Kwok S. C. M., Ledley F. D., DiLella A. G., Robson K. J. H., and Woo S. L. C. (1985) Nucleotide sequence of a full-length complementary DNA clone and amino acid sequence of human phenylalanine hydroxylase. *Biochemistry* **24**, 556–561.

Laemmli U. K. (1971) Cleavage of structural proteins during the assembly of the head of bacteriophage T4. *Nature* **227**, 680–683.

Lai C., Brow M. B., Nave K.-A., Noronha A. B., Quarles R. H., Bloom F. E., Milner R. J., and Sutcliffe J. G. (1987a) Two forms of 1B236/myelin-associated glycoprotein, a cell adhesion molecule for postnatal neural development, are produced by alternative splicing. *Proc. Natl. Acad. Sci. USA* **84**, 4337–4341.

Lai C., Watson J. B., Bloom F. E., Sutcliffe J. G., and Milner R. J. (1987b) Neural protein 1B236/myelin-associated glycoprotein (MAG) defines a subgroup of the immunoglobulin superfamily. *Immunol. Rev.* **100**, 127–149.

Land H., Schutz G., Schmale H., and Richter D. (1982) Nucleotide sequence of cloned cDNA encoding bovine arginine vasopressin-neurophysin II precursor. *Nature* **295**, 299–303.
Lee R. W. H. and Huttner W. B. (1983) Tyrosine-O-sulfated proteins of PC 12 cells and their sulfation by a tyrosylprotein sulfotransferase. *J. Biol. Chem.* **258**, 11326–11334.
Lemke G. and Axel R. (1985) Isolation and sequence of a cDNA encoding the major structural protein of peripheral myelin. *Cell* **40**, 501–508.
Lenoir D., Battenberg E. L. F., Kiel M., Bloom F. E., and Milner R. J. (1986) The brain specific gene 1B236 is expressed posnatally in the developing rat brain. *J. Neurosci.* **6**, 522–530.
Loh Y. P., Brownstein M. J., and Gainer H. (1984) Proteolysis in neuropeptide processing and other neural functions. *Ann. Rev. Neurosci.* **7**, 189–222.
Lucknow V. A. and Summers M. D. (1988) Trends in the development of baculovirus expression systems. *Bio/Technology* **6**, 47–55.
Masu Y., Nakayama K., Tamaki H., Harada Y., Kuno M., and Nakanishi S. (1987) cDNA cloning of bovine substance-K receptor through oocyte expression system. *Nature* **329**, 836–838.
Melton D. A. (1987) Translation of messenger RNA in injected frog oocytes. *Meth. Enzymol.* **152**, 288–296.
Milner R. J. and Sutcliffe J. G. (1983) Gene expression in rat brain. *Nucl. Acids Res.* **11**, 5497–5520.
Miller F. D., Naus C. C. G., Higgins G. A., Bloom F. E., and Milner R. J. (1987) Developmentally regulated rat brain mRNAs: molecular and anatomical characterization. *J. Neurosci.* **7**, 2433–2444.
Mishina M., Tobimatsu T., Imoto K., Tanaka K., Fujita Y., Fukuda K., Kurasaki M., Takahashi H., Morimoto Y., Hirose T., Inayama S., Takahashi T., Kuno M., and Numa S. (1985) Location of functional regions of acetylcholine receptor α-subunit by site-directed mutagenesis. *Nature* **313**, 364–369.
Nestler E. J. and Greengard P. (1984) *Protein Phosphorylation in the Nervous System.* Wiley, New York.
Neuberger A., Gottschalk A., Marshall R. D., and Spiro R. (1972) Carbohydrate-peptide linkages in glycoproteins and methods for their elucidation, in *The Glycoproteins: Their Composition, Structure and Function* (Gottschalk A., ed.), pp. 450–490, Elsevier/North-Holland, Amsterdam.
Nilsson B., Abrahmsen L., and Uhlen M. (1985) Immobilization and purification of enzymes with staphyloccal protein A fusion vectors. *EMBO J.* **4**, 1075–1080.

Prober J. M., Trainor G. L., Dam R. J., Hobbs F. W., Robertson C. W., Zagursky R. J., Cocuzza A. J., Jensen M. A., and Baumeister K. (1987) A system for rapid DNA sequencing with fluorescent chain-terminating dideoxynucleotides. *Science* **238**, 336–341.

Reeck G. R., de Haen C., Teller D. C., Doolottle R. F., Fitch W. M., Dickerson R. E., Chambon P., McLachlan A. D., Margoliash E., Jukes T. H., and Zuckerkandl E. (1987) "Homology" in proteins and nucleic acids: a terminology muddle and a way out of it. *Cell* **50**, 667.

Ruther U. and Muller-Hill B. (1983) Easy identification of cDNA clones. *EMBO J.* **2**, 1791–1794.

Sargan D. R., Gregory S. P., and Butterworth P. H. W. (1982) a possible novel interaction between the 3'-end of 18S ribosomal RNA and the 5'-leader sequence of many eukaryotic messenger RNAs. *FEBS Lett.* **147**, 133–136.

Shatzman A. R. and Rosenberg M. (1987) Expression, identification, and characterization of recombinant gene products in *Escherichia coli*. *Meth. Enzymol.* **152**, 661–673.

Shaw G. and Kamen R. (1986) A conserved AU sequence from the 3' untranslated region of GM-CSF mRNA mediates selective mRNA degradation. *Cell* **46**, 659–667.

Smith L. M., Sanders J. Z., Kaiser R. J., Hughes P., Dodd C., Connell C. R., Heiner C., Kent S. B. H., and Hood L. E. (1986) Fluorescence detection in automated DNA sequence analysis. *Nature* **321**, 674–679.

Staden R. (1982) An interactive graphics program for comparing and aligning nucleic acid and amino acid sequences. *Nucl. Acids Res.* **10**, 2951–2961.

Sutcliffe J. G. (1988) mRNA in the mammalian nervous system. *Ann. Rev. Neurosci.* **11**, 157–198.

Sutcliffe J. G., Milner R. J., and Bloom F. E. (1983a) Cellular localization and function of the proteins encoded by brain-specific mRNAs. *Cold Spring Harbor Symp. Quant. Biol.* **48**, 477–484.

Sutcliffe J. G., Milner R. J., Shinnick T. M., and Bloom F. E. (1983b) Identifying the protein products of brain-specific genes with antibodies to chemically synthesized peptides. *Cell* **33**, 671–682.

Sutcliffe J. G., Shinnick T. M., Green N., and Lerner R. A. (1983c) Antibodies that react with predetermined sites on proteins. *Science* **219**, 660–666.

Thomas P. (1980) Hybridization of denatured RNA and small DNA fragments transferred to nitrocellulose. *Proc. Natl. Acad. Sci. USA* **77**, 5201–5250.

Towbin H., Staehlin T., and Gordon J. (1979) Electrophoretic transfer of proteins from polyacrylamide gels to nitrocellulose. *Proc. Natl. Acad. Sci. USA* **77,** 5201–5205.

Travis G. H. and Sutcliffe J. G. (1988) Phenol emulsion-enhanced DNA-driven subtractive cDNA cloning: isolation of low abundance monkey cortex-specific mRNAs. *Proc. Natl. Acad. Sci. USA* **85,** 1696–1700.

Travis G. H., Naus C. G., Morrison J. H., Bloom F. E., and Sutcliffe J. G. (1987) Subtractive cDNA cloning and analysis of primate neocortex mRNAs with regionally-heterogeneous distributions. *Neuropharmacol.* **26,** 845–854.

von Heijne G. (1984) How signal sequences maintain cleavage specificity. *J. Mol. Biol.* **173,** 243–251.

von Heijne G. (1985) Signal sequences: the limits of variation. *J. Mol. Biol.* **184,** 99–105.

Wharton K. A., Johansen K. M., Xu T., and Artavanis-Tsakonas S. (1985) Nucleotide sequence from the neurogenic locus Notch implies a gene product that shares homology with proteins containing EGF-like repeats. *Cell* **43,** 567–581.

Williams A. F (1987) A year in the life of the immunoglobulin superfamily. *Immunol. Today* **8,** 298–303.

Young R. and Davis R. (1983) Efficient isolation of genes by using antibody probes. *Proc. Natl. Acad. Sci. USA* **80,** 1194.

Isolation and Structure Determination of Genes

Wendy B. Macklin and Celia W. Campagnoni

1. Introduction

The study of genes that are expressed only, or with unique effects, in the nervous system has become important from both a molecular biological and a clinical perspective. From knowledge of the structure of a gene and its exons, it is possible to identify

1. the sequence of the mRNA produced by that gene
2. mechanisms of gene expression such as steroid regulation, or alternative mRNA splicing
3. evolutionary relationships of genes or
4. mutations that alter expression of a gene, causing disease.

Analysis of genomic DNA sequences allows identification of promoter elements or other regulators, such as enhancers, which can be as much as 10 Kb away from the exons of the gene. These gene regulators are at least as important as the structural elements of the gene for the normal function of the gene. From a compilation of DNA sequences of many genes, several short DNA sequences have been identified, which appear to be involved in gene regulation. For example, a 14-nucleotide consensus sequence has been identified in genes that respond to glucocorticoids (Karin et al., 1984) and a seven-nucleotide consensus sequence appears to be common to genes that respond to metal activators (Stuart et al., 1985). As more genes are sequenced and the regulatory features of those genes are demonstrated, many more such consensus sequences will become apparent, hopefully providing information on the regulation of newly identified genes.

Another element that regulates gene expression is the generation of multiple mRNAs from a single gene through alternative splicing. The structure of a gene, in particular the exon/intron splice positions, must be established in order to confirm that alternative splicing can occur, and to understand the regulation of the splicing mechanism. The expression of some alternatively spliced mRNAs is developmentally regulated, and the regulation of the splicing process may be important for normal brain function. The myelin basic protein gene is spliced to produce at least five different mRNAs, and the ratio of these mRNAs changes during brain development (Campagnoni et al., 1978; Carson et al., 1983). The calcitonin gene is spliced to produce calcitonin gene-related peptide (CGRP) in neurons, but calcitonin in thyroid C cells (Leff et al., 1987). This results from the presence of a factor(s) that determines the specific splicing pathway in neurons and presumably thyroid cells.

Relationships among different genes can be demonstrated by analysis of both sequence data and exon–intron relationships. A distant relationship has been demonstrated for the prolactin and growth hormone genes (Niall et al., 1971; Cooke and Baxter, 1982), and for the transferrin and ovotransferrin genes (Cochet et al., 1979; Park et al., 1985) on the basis of their exon structures. Evidence that the transferrin gene itself arose by gene duplication was provided by characterizing isolated transferrin genomic clones. It was evident that, in the two halves of the gene, sets of exons were homologous in both size and sequence (Park et al., 1985).

Two devastating neurodegenerative disorders, Huntington's disease (HD) and Duchenne's muscular dystrophy (DMD), are classical examples of the value of investigating gene structure. HD is a highly penetrant autosomal dominant disorder that produces the loss of certain striatal neurons (Martin and Gusella, 1986). Although the molecular basis for the cell loss in HD is totally unknown, the genetic alternation has been mapped on human chromosome 4 (Gilliam et al., 1987). The HD gene has yet to be isolated, but once it is isolated, investigating its structure and function will enhance our understanding of the cause of HD.

Another example of the clinical value of studying gene structure is DMD, an X-linked neuromuscular disorder. The DMD gene was recently isolated; it is over two million bases in length, and it encodes an mRNA of approximately 14,000 bases (Koenig et al., 1987). Apparently, multiple alterations of the DMD gene have occurred within the human population. Certain areas of the DMD gene are particularly sensitive to major structural changes, e.g., deletions, and the specific alterations can vary among different individuals. With the identification of dystrophin as the protein product of the DMD gene (Hoffman et al., 1987), and the demonstration that the loss of this protein in DMD patients primarily affects fast muscle fibers (Webster et al., 1988), clinical therapies may become possible.

In order to investigate the regulation of the expression of a gene, it is essential to have cloned genomic DNA containing the gene. This chapter will be devoted to protocols for preparation of genomic libraries, isolation of cloned genomic DNA, mapping of genomic DNA clones, and subcloning and sequencing of genomic DNA. The protocols discussed here have been used successfully in our laboratories, but of necessity their discussion has been somewhat abbreviated. For a more extensive discussion, refer to Maniatis et al. (1982), Davis et al. (1986), or Berger and Kimmel (1987).

2. Generation of Genomic Libraries

2.1. Preparation of High Mol Wt Genomic DNA

We have successfully used the following procedure in our laboratory for the isolation of DNA. It avoids precipitating the DNA with all the attendant difficulties in getting high mol wt DNA back into solution. It is extremely important that the DNA to be used for preparing a genomic DNA library have a high mol wt. Thus, the sample should be handled carefully to prevent mechanical shearing of the DNA.

1. Prechill a 12–37 mL stainless-steel mini-container for a Waring blender by grinding small pieces of dry ice in it. Blend in 20-s pulses, releasing the pressure after each pulse.
2. Add more dry ice and a few hunks of frozen liver, and blend until 2 g of liver have been pulverized. The liver must remain frozen at all times, but ideally, by the last homogenization, very little dry ice should remain in the blender.
3. Transfer the tissue to a 50-mL screwcap centrifuge tube with 10 mL of extraction buffer: 50 mM Tris pH 8, 100 mM NaC1, 50 mM EDTA
4. Rinse the blender container with 10 mL extraction buffer.

5. Add 1 mL 10% SDS, 100 μL proteinase K (10 mg/mL), and incubate 2 h at 37°C.
6. Extract briefly with an equal volume of UNC-phenol by gently inverting the tube. (UNC-phenol: 454 g phenol, 100 mL 2M Tris pH 7.6, 130 mL H_2O, 25 mL m-cresol, 1 mL β-mercaptoethanol, 500 g 8-hydroxyquinoline). Centrifuge and remove the aqueous phase.
7. Extract with an equal vol of UNC-phenol:$CHCl_3$, 1:1.
8. Extract with an equal vol of $CHCl_3$.
9. Dialyze overnight (ON) against 10 mM Tris pH 8, 1 mM EDTA.
10. Add 2M Tris pH 8 to a final concentration of 20 mM and 0.5M EDTA pH 8 to 10 mM.
11. Digest with 100 μL of 10 mg/mL DNase-free RNase for 1 h at 37°C. DNase-free RNase can be prepared by heating a solution of RNase in 20 mM sodium acetate pH 6 in a beaker of boiling water for 10 min, and then allowing the tube and the beaker of water to return to room temperature slowly.
12. Digest for 2 h at 37°C with 100 μL of proteinase K (10 mg/mL).
13. Extract with an equal vol of UNC-Phenol.
14. Concentrate the DNA by extracting successively with one or two vol of butanol until the vol of the aqueous phase is reduced to 5 mL.
15. Dialyze against 10 mM Tris pH 7.6, 0.1 mM EDTA.
16. Dilute 50 μL to 1 mL, measure the absorbance at 260 nm, and multiply by 40 μg/A_{260} U. In general, 0.8–1.4 mg DNA/g of liver should be obtained.
17. Check that the DNA is of higher mol wt than λ DNA by electrophoresis on a 0.4% agarose gel in Howley's buffer. 40X Howley's buffer: 193.6 g Tris, 108.9 g NaAc-3H_2O, 14.8 g Na_2EDTA-2H_2O adjusted to pH 7.2 with glacial acetic acid.

2.2. Preparation of Genomic Library

Once high mol wt genomic DNA is prepared, it can be used to generate a genomic library. Currently, one of the most commonly used λ vectors is EMBL3/4 (Frischauf et al., 1983), although many others, such as the Charon vectors (Blattner et al., 1977), have also been used. Our laboratories have used the EMBL vectors and λ Jl because of their ability to accept up to 23 kb DNA, and the presence of a polylinker, which simplifies production and characterization of the clones. In addition, they include a χ element, which makes the plaque size more uniform (Stahl et al., 1975).

When larger fragments of DNA are to be cloned, other vectors are utilized. Cosmids have been engineered to contain minimal viral or plasmid DNA, so that they can accept up to 45 kb insert DNA (Collins and Hohn, 1978). In general, cosmid libraries are more difficult to store and to screen than λ libraries. In addition, because of the method of library construction, occasionally two genomic fragments that are unlinked in the genome become linked and inserted in the cosmid. However, methods are becoming available to prevent such problems (Ish-Horowitz and Burke, 1981; Lau and Kan, 1983). A more extensive discussion of the benefits and disadvantages of the cosmid library is presented elsewhere (Maniatis et al., 1982; DiLella and Woo, 1987).

A recent report indicates the usefulness of another vector, yeast artificial chromosomes, for cloning foreign DNA up to several hundred kb in length (Burke et al, 1987). The vector is a plasmid, containing a cloning site, and several important DNA segments that permit inserted DNA to be replicated in yeast as artificial chromosomes. Clearly, this vector and cosmid vectors will be increasingly important for characterizing genomic DNA, although currently the λ replacement vectors are used much more extensively.

In order to clone genomic DNA into the λ vectors, the genomic DNA must be of the appropriate size (approximately 20 kb). Standard protocols for generating these fragments utilize partial digestion with *Mbo* I. This enzyme, which recognizes a four-nucleotide sequence, generates DNA fragments with *Bam* HI-compatible ends for ligation into the λ polylinker. Because of the high frequency of *Mbo* I sites within DNA, a limited partial digest will generate large, overlapping fragments, which is essential for cloning genes greater than 10–15 kb in length. A detailed protocol for cloning *Mbo* I-digested genomic DNA into the *Bam* HI site of EMBL3 is provided here as a general method. EMBL3 is particularly useful, since the insert is excised from the vector with *Sal* I. Since *Sal* I sites are underrepresented in the mammalian genome, many inserts will have no internal *Sal* I sites to complicate subsequent insert mapping (Frischauf et al., 1987). However, *Sal* I is more expensive than *Eco* RI (the enzyme used to excise inserts from EMBL4), and its activity is very sensitive to impurities in DNA preparations.

2.2.1. Preparation of Insert DNA

1. Conduct test digestions of the high mol wt genomic DNA to establish optimal conditions for preparing 15–20 kb DNA fragments. Initially, digest 10 μg genomic DNA with 0.1, 1.0, 10, or 20 U of *Mbo* I restriction enzyme in a final volume of 100 μL at 37°C.
2. Remove 15 μL aliquots at 5, 10, and 20 min, add 1 μL 0.5 M EDTA and analyze each by electrophoresis in 0.6% agarose TBE gels (TBE: 0.05 M Tris, 0.05M boric acid, 1 mM EDTA), with appropriate mol wt markers.
3. Once optimal conditions have been identified for the partial digest, scale up the reaction 20-fold. Digest 200 μg genomic DNA, and then precipitate the DNA with 2 vol of ethanol at −20°C for 3–16 h.
4. Resuspend the resulting pellet in 1 mL TE (10 mM Tris, pH 7.4, 1 mM EDTA), and fractionate it by gradient centrifugation. (Some investigators find that DNA purified by gel electrophoresis contains impurities from the agarose, which inhibit the ligation reaction.) For gradient purification, use 38 mL ultracentrifuge rotor tubes to pour two 5–20% potassium acetate gradients in THE (10 mM Tris, pH 7.4, 5 mM EDTA), with the heavy solution displacing the light solution.
5. Layer 0.5-mL samples on top of each gradient, and centrifuge in an SW28 rotor at 20,000 rpm, 22°C for 16 h.
6. Collect 1-mL fractions and analyze 30 μL test aliquots by electrophoresis on a 0.7% agarose TBE gel.
7. Pool those fractions containing 15–20 kb DNA fragments and ethanol precipitate the DNA.
8. Resuspend the final pellet in 0.5 mL TE.

2.2.2. Ligation and Packaging of DNA

1. Set up test ligations using 3 μg EMBL3 arms (Promega, Stratagene) and 1.5 μg insert DNA (or 3-μg

arms and 1-µg insert DNA), 1 U T4 DNA ligase, 1 µL 10× ligase buffer (10× ligase buffer: 250 mM Tris pH 7.8, 100 mM MgCl$_2$, 10 mM dithiothreitol, 4 mM ATP). ATP should be prepared as a fresh solution, neutralized, and added to buffer. It is possible to prepare the λ arms in the laboratory, using protocols described by Maniatis et al. (1982) or Davis et al. (1986).
2. Increase reaction vol to 10 µL and ligate ON at 16°C.
3. Package the ligated DNA, using a commercially available packaging extract (e.g., Packagene: Promega Biotech, or Gigapack: Stratagene). Add each ligation reaction to a tube of packaging extract, and incubate at 22°C for 2 h.
4. Add 0.5 mL phage dilution buffer to each packaged ligation (phage dilution buffer: 0.1M NaCl, 0.02M Tris, pH 7.4, 0.01M MgCl$_2$).
5. Prepare plating cells by growing bacteria (LE392) in LB broth (LB broth: 10 g bacto-tryptone, 5 g bacto-yeast extract, 10 g NaCl/L, pH 7.4) containing 10 mM MgS0$_4$, to an O.D.$_{600}$ of 1.0. Centrifuge cells, and resuspend pellet in one-tenth the initial vol of cold 10 mM MgS0$_4$ or SM buffer (50 mM Tris pH 7.5; 8 mM MgS0$_4$, 0.1M NaCl, 0.01% gelatin). MgS0$_4$ helps to stabilize the phage.
6. Plate three concentrations of packaged ligation mix with plating cells. Prepare a 1:100 dilution of each packaged ligation mix using phage dilution buffer. Add 1 µL and 10 µL of the dilute packaged ligation mix and 1 µL of the undiluted packaged ligation mix to 100 µL of plating cells; incubate for 20 min at 37°C.
7. Add 2.5 mL melted LB top agar (LB broth containing 0.8% agar, 45–50°C) containing 10 mM MgS0$_4$, mix, and pour over LB plates (1.5% agar). Swirl the mixture gently, taking care to avoid introducing bubbles into the top agar. Pour the suspension onto LB plates. Tilt the plates to spread the top agar evenly, and let the agar solidify for at least 10 min at room temperature. Invert the plates and incubate ON at 37°C.

8. Use the information from the test ligations and packagings for optimal preparation of the full library. Ligate and package enough DNA to generate a library with a base of approximately 10^6 members. It is advisable to ligate and package several small samples, rather than combining ligation reactions into a larger vol. Pool the separate packaging reactions after diluting each with 0.5 mL phage dilution buffer.
9. Titer the library to determine its actual base. Suspend 5 mL stationary phase cells in 5 mL 10mM MgSO$_4$ or SM buffer. Make a series of 10-fold dilutions of the library in 10 mM MgSO$_4$ or SM buffer. The typical titer for a library is a property of the vector phage. It can range from 10^8–10^{11} plaque-forming U (pfu)/mL. Mix 100 µL of cells with phage dilution aliquots (vol approximated from the earlier test reactions), and incubate for 20 min at 37°C. Add 2.5 mL of LB top agar. Spread top agar and let solidify as above. Incubate 6–12 h for an EMBL 3/4 library. Phage clones appear as pinpoint plaques or clear areas on the bacterial lawn. Do not allow phage plaques to touch, as recombination can occur. Calculate the titer as the pfu/mL of the library stock.

There are advantages and disadvantages to amplifying the library at this point. The advantage is that the library can be replated and rescreened numerous times. Thus, if a mouse genomic DNA library will be screened for several different clones over several years, it is likely that an amplified library will be needed. There is a 94% chance that any given sequence of DNA in a mammalian genome, complexity 3×10^9, will be represented in a screen of 500,000 members of a library in which the average insert size is 17 kb (Clarke and Carbon, 1976). When screening an amplified library, it is reasonable to screen 5–7 times that amount to be certain of screening every member of the library.

One disadvantage of amplification is that certain clones may be under- or overrepresented in the amplified library, depending on the ability of the clones to replicate (Wyman et al., 1986). Some genomic fragments, for example, palindromic sequences, will inhibit phage replication (Leach and Stahl, 1983), which will make it extremely difficult to isolate those clones from an amplified library.

Other clones may be overrepresented in the library, and many isolated clones will be identical. Another disadvantage of amplification, when using EMBL3/4 or other λ replacement vectors in a RecA$^+$ bacterial strain, is an increased risk of recombination, which could generate artifacts in the subsequently isolated clones. LE392 is a commonly used RecA$^+$ bacterial strain, which has been used to amplify many libraries with few recombination problems. However, if desired, λ replacement libraries can be grown in RecB-, Rec C-, sbc- cells such as CE201, which should reduce problems of underrepresentation and recombination (Wyman and Wertman, 1987).

If the library is to be amplified, it may be preferable to amplify only a part of the library. We have generally amplified the libraries as in Davis et al. (1986). *It is crucial from this point onward that no cloned DNA is inadvertently introduced into the library.* Some cloned plasmid or phage contaminants can replicate faster than the library phage and will replace much of the library. Thus, automatic pipeters, which are potentially significant sources of contamination, should be cleaned regularly.

To screen the library, plate either the entire unamplified library or 50–100 plates of the amplified library, at 10,000–200,000 plaques/plate *(see below).*

3. Isolation of Genomic Clones

3.1. Preparation of Probe

A genomic DNA library is most efficiently screened using a complete or almost complete cDNA for the protein of interest, or the cloned DNA from the same gene in another species. Oligomers synthesized from the protein sequence can then be used to identify any missing exons (Takahashi et al., 1985). It is not unusual for the genes of even relatively small proteins, for instance the myelin basic protein (204 amino acids) or the myelin proteolipid protein (276 amino acids) to be encoded in many small exons (33–200 bases) spread over 15–35 kb on the chromosome. Each exon must be identified, subcloned, and sequenced; thus, DNA hybridizing to each of them must be available for the gene analysis.

Before they can be used, some further engineering of the cloned cDNA or genomic probes may be necessary. Phage clones should be subcloned into plasmids. A 1–5 kb insert represents a much larger percentage of the total DNA when inserted in 3–5-kb plasmid as compared to a 50-kb phage. Preparation of insert from plasmids requires less DNA and less enzyme than from phage. Because some untranslated regions of cDNA and some genomic introns contain DNA elements that make them unsuitable as probes (for example, highly reiterated DNA), each potential probe should be tested on a portion of the library before it is used for an entire screen.

3.1.1. Isolation from a Gel

In order to isolate the probe, it must be excised from the vector and separated from it by electrophoresis. Plasmid DNA is digested for 2 h with 2 U of enzyme/µg of DNA, at a concentration of 0.1 µg DNA/µL using the buffer and temperature recommendations of the enzyme supplier. Samples are then applied to electrophoresis gels.

There is a wide range of techniques for the isolation of insert from the digested plasmid: electrophoresis in acrylamide or agarose gels, followed by purification of the DNA from excised bands by simple diffusion (Maxam and Gilbert, 1980) or electroelution (Wu et al., 1976; Smith, 1980; Southern, 1975), electrophoresis on low-melt agarose (Wieslander, 1979), excision of the bands and extraction of the DNA from the liquified agarose, separation on agarose gels followed by electrophoresis of the bands onto DEAE-cellulose paper (Dretzen et al., 1981; Winberg and Hammarskjold, 1980) or DEAE-nylon membrane, or electrophoresis on disulfide-acrylamide gels, and isolation of the insert by solubilization of the gel in β-mercaptoethanol and DEAE cellulose chromatography (Hansen, 1981). Our laboratories have most frequently used either the disulfide gel procedure or electroelution. The disulfide gel technique gives us the best recoveries—around 90% for fragments 2 kb or less, falling to about 50% for fragments of 8 kb—and it results in fewer problems with cross-contamination of fragments; however, disulfide gels can be tricky. The polymerization temperature is critical. Gels that polymerize too slowly form products that do not dissolve in mercaptoethanol and subsequently clog the DEAE cellulose column. The separation of high mol wt fragments can require extensive electrophoresis time. Although the recoveries from electroelution are only about 50%, both the time required

Isolation and Structure of Genes

for the separation of large fragments and for the work-up of the insert are considerably shorter than with the disulfide procedure.

3.1.1.1. DISULFIDE GELS. Utilization of disulfide gels is detailed in the following protocol:

1. Equilibrate a gel apparatus at 37–42°C for at least 1 h before pouring the gel. We use a 13 × 36 × 0.15-cm gel.
2. Mix 50 mL 10% acrylamide stock, 44 mL water, 5 mL 20× TBE for each 100 mL of gel solution, and degas the mixture. Acrylamide stock: 95.5 g acrylamide, 4.5 g bisacrylocysteine/L.
3. Heat the solution to 45°C, and add 1 mL freshly prepared ammonium persulfate (100 mg/mL) for each 100 mL of gel solution.
4. Cool the mixture to 43°C; add TEMED (0.25 mL/100 ml) and quickly pour the gel. Work fast, because polymerization should occur within 2 min.
5. Incubate the gel for 1 h at 37–42°C.
6. Rinse the wells with the running buffer, TBE. Dissolve sample in 20 µL 10mM Tris pH 7.6; add 6 parts dye mixture (5 mL glycerol, 2 mL 2% xylene cyanol blue, 2 mL 2% bromophenol blue, 0.5 mL 20% SDS, and 0.5 mL water), and load on the gel. Apply 4–8 mA constant current until the dye markers have moved the appropriate distance to separate the desired insert from other DNA fragments. It is possible to estimate the position of each fragment from the graph in Fig. 1. For example, at this acrylamide concentration, fragments of 82 bases will migrate half the distance moved by bromophenol blue, whereas fragments of 540 bases will migrate with xylene cyanol blue. For much larger fragments, successive loads of dye may need to be added to short gels, since the xylene cyanol runs off the gel, and the distance the dye has migrated is measured as the sum of the distance successive loadings of xylene cyanol have migrated on the gel. The following information may be useful. When the xylene cyanol migrates 39 cm, 1-kb fragments will migrate

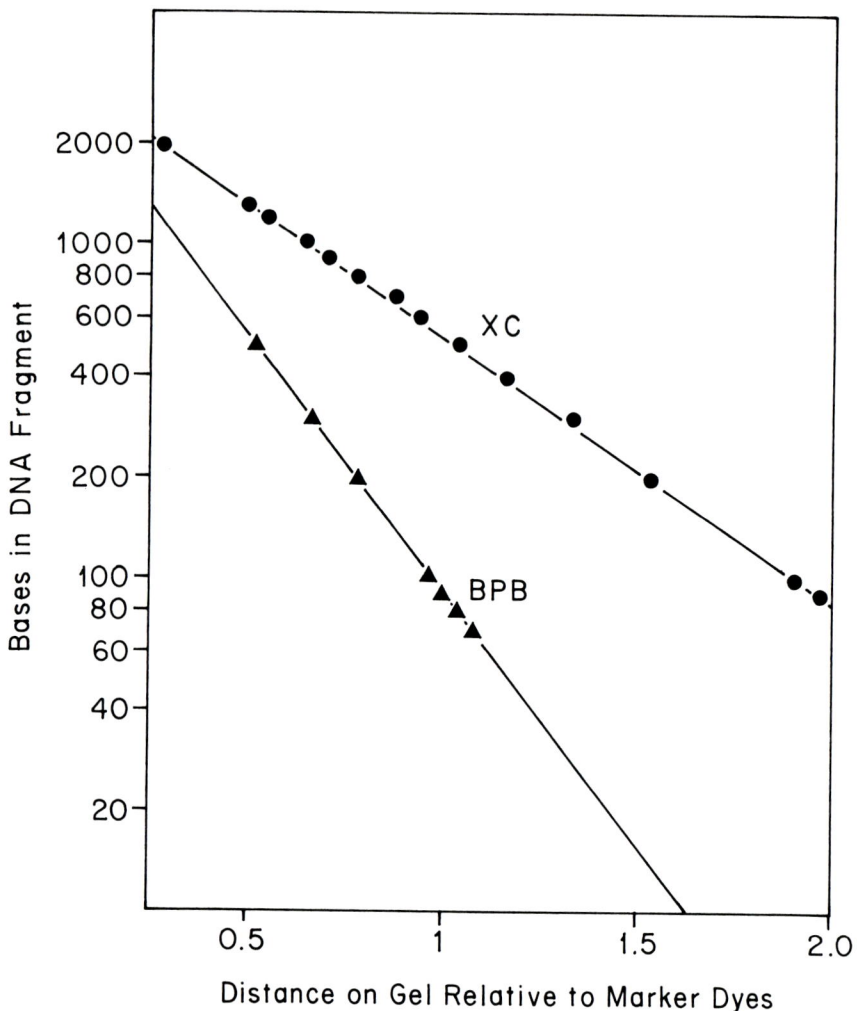

Fig. 1. Mobility of DNA fragments on a disulfide gel compared to the migration of marker dyes. The line defined by triangles shows the mobility relative to bromophenol blue, that defined by circles, the mobility relative to xylene cyanol blue.

approximately 19 cm; 4-kb fragments, 7 cm; and 7.7 kb fragments, 5 cm.

7. Remove one of the gel plates, leaving the gel, which will be very fluid, attached to the other. Keep gloves and the gel wet to prevent tearing the gel. Stain the gel for 10 min in dilute ethidium bromide solution, and rinse it with water. Cover the gel plate with plastic wrap, invert the sandwich, and release the gel onto the plastic wrap for viewing on a transilluminator.
8. When excising the insert band, minimize the size of the gel piece, if necessary cutting off the trailing edge of the band. Gel pieces can be stored at 4°C for several weeks.
9. Pour a 1cm DEAE cellulose column (Whatman DE23) in a siliconized, plugged Pasteur pipet, and equilibrate with TBE.
10. Dissolve the gel fragment in either 5 mL TBE + 100 µL β-mercaptoethanol (if the gel piece is approximately 0.5 × 15 × 0.15 cm) or 2.5 mL TBE and 75 µL mercaptoethanol for smaller amounts of gel. Immediately apply the dissoved gel to the ion exchange column.
11. Wash the column with two vol of TBE, and one vol of low-salt buffer: 10 mM Tris pH 7.6, 0.2M NaCl.
12. Break the tip of the pipet off just below the glass-wool plug. Apply 50 µL high-salt buffer to the column, and discard the eluate. High-salt buffer: 10 mM Tris pH 7.6, 1.0M NaCl, 6.5M urea.
13. Elute the DNA into two Eppendorf tubes with two 500 µL aliquots of high-salt buffer. Add 1 mL of ethanol to each, incubate at –70°C for 30 min, and then centrifuge 10 min at 4°C.
14. Add 100 µL 0.3M sodium acetate pH 6 to the pellet and extract with an equal volume of UNC-phenol, vortexing for at least 1 min. This vortexing step is important, since the DNA is released from the column in a water-insoluble form and will remain in the phenol phase for a long time. Large fragments are hard to release by phenol extraction, which is a major cause for a low yield in this technique.

15. Reextract the phenol phase with 100 μL sodium acetate pH 6, again vortexing for at least 1 min.
16. Combine the aqueous phases, and extract them with 400 μL diethyl ether.
17. Precipitate the DNA with 600 μL of ethanol at –70°C for 30 min.
18. Centrifuge for 10 min at 4°C. Remove the supernatant, and carefully layer 1 mL of room temperature ethanol over the pellet. Centrifuge 10 min at 4°C. Remove the supernatant, and dry the pellet *in vacuo* for 10 min. Dissolve the material in 10 m*M* Tris pH 7.6, and run a small aliquot on an agarose gel along with known concentrations of DNA to estimate the concentration.

3.1.1.2. ELECTROELUTION. Apparatuses for electroelution range from a simple homemade device described by Davis et al. (1986) to a wide variety of commercial designs. In each, following electroelution, the DNA solution is extracted with an equal vol of $CHCl_3$: UNC-phenol, 1:1 and then with 2 vol of diethyl ether. The DNA is precipitated with 3 vol of ethanol at –70 or –20°C, ON.

3.1.2. Labeling of Fragments

Our laboratories label synthetic oligonucleotides or very small double-stranded DNA probes by kinasing. Larger probes are labeled by either random priming or nick translation. We generally obtain specific activities of 1×10^8 cpm/μg for kinased oligomers and nick-translated fragments, and 1×10^9 cpm/μg for those labeled by random priming.

3.1.2.1. KINASING. In this reaction, the phosphate from ATP is transferred to the 5' OH of the nucleic acid. Any 5' phosphate groups must be removed before DNA fragments can be labeled with T4 polynucleotide kinase by the procedure described in section 4.4. The kinasing itself is described in detail by Maxam and Gilbert (1980).

3.1.2.2. NICK TRANSLATION. During nick translation, the DNA is nicked with DNase I, and then DNA polymerase I removes nucleotides on the 5' side of the nick, while adding ^{32}P-labeled nucleotides to the 3' side. This results in movement of the nick along the DNA and the synthesis of a short labeled molecule (Rigby et al., 1977).

Mix approximately 200 ng insert DNA with 50 μCi α^{32}P-dATP (>600 Ci/mmole), 3 μL 10× NT buffer, 3 μl 10× A° mix, 0.3 ng DNase I (Worthington, DPFF, >2000 U/mg), 1 U DNA Pol I, and make up to 30 μL with water. (1× NT buffer: 25 mM Tris, pH 7.8, 50 mM NaCl, 10 mM MgCl$_2$, 100 μg/mL bovine serum albumin [BSA], 1 mM dithiothreitol. 10× A° mix: 0.5 mM dCTP, 0.5 mM dGTP, 0.5 mM dTTP. Any labeled deoxynucleotide can be used, with the appropriate N° mix.) Incubate reaction mix at 17°C for approximately 1 h. Labeled probe can be purified from unincorporated isotope using ethanol precipitation, disposable mini-spin columns (Boehringer Mannheim or Worthington), or Sephadex G50 column chromatography.

3.1.2.3. RANDOM PRIME LABELING. In the random prime labeling procedure, random oligonucleotides are used as primers to synthesize radiolabeled copies of the probe (Feinberg and Vogelstein, 1983). Prepare LS buffer by mixing 5 μL 1M HEPES pH 6.6, 5 μL of a deoxytrinucleotide mixture (DTM), and 1.4 μL oligodeoxyribonucleotides (hexamers-Pharmacia) at a concentration of 90 OD U/mL in 1 mM Tris pH 7.5, 1 mM EDTA. DTM: 100 μM each of dATP, dGTP, dTTP, and dCTP in 250 mM Tris pH 8, 25 mM MgCl$_2$ and 50 mM β-mercaptoethanol. Heat a tightly sealed tube containing 63 ng of DNA in 6.3 μL of water in a boiling water bath for 2 min, and then plunge the tube into an ice bath. Add 11.4 μL LS buffer, 1 μL nuclease-free BSA (10 mg/mL), 50 μCi of α^{32}P-dCTP (3000 Ci/mmole) and 1 μL large fragment of DNA polymerase 1 (2.5 U/μL); mix the solution by gently pipetting it up and down several times. Incubate the reaction at room temperature for 6–18 h. Isolate the labeled DNA by any of the procedures described in nick translation.

3.2. Clone Purification

Screening the library will require at least three rounds of purification. The first round examines 500,000 members of the library at high density, 10,000 pfu/plate. In the second round, phage from a positive region of the plate are examined at a density of 2000–5000 pfu in order to isolate positive clones cleanly from the rest of the library. Round 3, the array, checks the identity of the purified clones, and provides a source of phage for mapping and subcloning experiments. The first step in working with any library is to determine the concentration of active phage particles in that library accurately. This titer can fall by as much as 10-fold in a year, so it is important to measure it each time just before the library is screened. The titering procedure is described in section 2.2.2.

3.2.1. Round 1: Initial Screen

1. Prepare 50–100 plates, and let them stand at RT or 37°C until any moisture has evaporated from the lids and the surface of the agar is dry. If the agar is wet, the top agar will lift off with the membrane during the blotting procedure.
2. Determine the maximum number of phage that will yield discrete, not confluent plaques on a 100-mm plate. This concentration will vary from about 10,000 to 20,000 pfu/plate depending on the vector.
3. For each batch of 10 plates, mix 1 mL host cells in SM buffer with 1–500 µL library in a sterile 50-mL tube, and incubate for 20 min at 37°C.
4. Dilute the phage-cell incubation to 31 mL with 45°C top agar, and spread 3 mL aliquots on 10 plates. Let the top agar harden. Then invert the plates, and incubate them at 37°C ON.
5. Chill the plates at 4°C for at least 3 h. Using a ball-point pen, number 50–100 nitrocellulose or nylon filters to correspond to the numbers on the bottom of the plates. Carefully lower the filter onto the agar. The filter cannot be moved once it has touched the plate without smearing the blot.
6. Allow the membrane to wet. Then mark its position on the plate by plunging a sterile 18-gage needle through the filter and into the agar in 3–4 places. This is absolutely essential for subsequent localization of plaques.
7. Carefully lift the membrane from the plate, and allow it to air dry, phage side up, on a sheet of filter paper.
8. Follow the manufacturer's instructions for the hybridization membrane you are using. For nitrocellulose, the phage are lysed by carefully floating the filter until it wets in a tray of $0.2M$ NaOH, $1.5M$ NaCl (500 mL/50 filters) or by very slowly lowering the membrane into this solution. Use this technique

for wetting the filter whenever a dry membrane is introduced to a solution. It is important to avoid abruptly immersing the filter and trapping air bubbles between the two wet surfaces of the membrane.
9. Agitate the filter for 30 s in the NaOH/NaCl solution, for 30 s in 0.4M Tris pH 7.6, 2 × SSC, and for 30 s in 2× SSC. 20× SSC: 175.3 g NaCl, 88.2 g sodium citrate ·2H$_2$0/L.
10. Air dry the filters, phage side up. Then layer them between paper towels and dry for 2 h at 80°C in a vacuum oven.

3.2.1.1. HYBRIDIZATION.
1. Incubate the filters for 1 h at 70°C in 500 mL/50 filters of 4× Denhardt's solution, 6× SSC. 100× Denhardt's solution: 20 g ficoll, 20 g polyvinylpyrrilidone, 20 g BSA fraction V/L.
2. Blot the filters between paper towels, but do not let the filters dry. Place the filters in a plastic sealing bag, 16–17 to a bag, and seal the sides of the bag as close to the filters as possible.
3. Pour in 16–17 mL of probe solution, squeeze out the air bubbles from between the filters, and seal the bag.
   Probe solution: Add the following reagents in the order in they are listed.
   0.5–0.6 µg of probe—0.5× 10^8 cpm
   40 µl denatured herring sperm or salmon sperm DNA (10 mg/mL) *
   1 mL 1M NaOH **
   0.25 mL 2M Tris pH 7.6 **
   1 mL 1M HCl **
   H$_2$0 (to vol)
   2 mL 100× Denhardt's solution
   15 mL 20× SSC
   0.25 mL 20% sodium dodecyl sulfate (SDS)
   total 50 mL
   * Herring sperm or salmon sperm DNA: dissolve the DNA in water and shear it to 1–2 kb lengths by passing it through a French press or an 18-gage needle several times. Denature

the DNA by boiling the solution for 10 min. Store in aliquots at –20°C.
** The final pH of the probe solution must be between 7 and 7.6. Adjust the volumes of these reagents accordingly.
4. Place the bag upright in a 70°C water bath for 10 min, and then squeeze any air bubbles that may have formed between the filters to the top of the bag. Incubate ON or at least 16 h at 70°C.

3.2.1.2. WASH CONDITIONS. The wash conditions will vary with the probe. Our laboratories usually begin with nonstringent conditions (two 30-min washes at 42°C in 2× SSC– 0.1% SDS) and only proceed to two 30-min washes at 55°C in 0.1× SSC, 0.1% SDS if the backgound is high.

1. Rinse the stacks of filters in 500 mL of wash solution. Blot each filter between paper towels, but do not let the filters dry.
2. Place the filters in a covered box with 500 mL wash solution, and incubate in a water bath, preferably with gentle agitation. Begin timing the wash when the solution within the box itself reaches the desired temperature.
3. Blot the filters between paper towels, and wash them again for 30 min in 500 mL of wash solution. Blotting filters each time between washes significantly reduces the background.
4. After the final wash, blot between paper towels. Attach the filters to a sheet of 3-mm filter paper using small pieces of plastic tape on the very edge of the filter.
5. Cover the filters and the filter paper with plastic wrap. Enclose the package in a photographic cassette with Kodak X-Omat AR film and a DuPont Lightning-Plus intensifying screen. Place a weight on the cassette to be certain that the film makes firm, even contact with the filters. Expose the film for 4–48 h at –70°C. Small plaques require longer exposure times than large ones.

Isolation and Structure of Genes

A positive clone often gives a slightly comet-shaped spot, the tail being produced as a portion of the clone is swept along when the filter wets. In extreme cases, the clone can even appear as an exclamation point where the clone itself is located in the dot rather than the slash portion of the exclamation mark. Perfectly circular spots are often, but not always, false positives.

3.2.1.3. TROUBLESHOOTING.   If the entire library appears to react with the probe, check that:

1. The probe does not contain a repetitive DNA sequence.
2. The probe does not contain a multi-purpose linker region, which is present in the library vector.
3. The host bacteria do not contain a plasmid that could be reacting with a small amount of plasmid DNA contaminating the probe preparation.
4. A suitable DNA hybridization membrane has been used. Batches of a nitrocellulose-type membrane that work well for Western blotting can have a very high background in Southern blots.
5. The hybridization membrane has not aged too much.

3.2.2. Round 2: Enrichment

Areas the diameter of a pencil and centered on the positive plaques are excised from the library plates, placed in 1 mL SM buffer, and approximately 10 µL of $CHCl_3$ is added to inhibit bacterial growth. The approximate titer of the phage suspension is then determined.

3.2.2.1. DIME TITERS.
1. Make a sequential series of 10-fold dilutions of the phage from 10^{-3} to 10^{-9}.
2. Mix 100 µL of an ON culture of host bacteria in SM with 3 mL of 45°C top agar, and spread this suspension on a dry plate. (The plate must be dry or the phage will escape from the "dime" in which it was applied and make the plates unreadable.)
3. Prepare a rough template fitting six dime-sized circles into a circle the size of a petri dish.
4. Allow the top agar to solidify, place the plate over the template, and then gently layer 10 µL of each

phage dilution onto the plate in the center of the dimes. They should form almost dime-sized puddles.
5. Carefully transfer the plate to a 37°C incubator, and incubate right-side up ON. At least one of these dilutions should give a readable number of plaques.

3.2.2.2. PLATING THE PHAGE FOR ROUND 2.   Plate two concentrations of phage for each positive clone, one that should yield approximately 2000 pfu on a plate and the other about 5000 pfu. For blotting, choose the plate that has the greatest number of cleanly separated plaques. Phage can diffuse some distance in top agar, so avoid plates where the plaques are separated by less than a few millimeters. Blot as described in round 1.

3.2.3. Round 3: Array

1. Prepare a pair of templates, each containing an identical square of 64 spots, eight rows across and eight rows down, which can be centered in a 100-mm petri dish-sized circle.
2. Mix 3 mL of 45°C top agar per plate with 100 µL of cells in SM and spread the suspension on a dry plate. Center the petri dishes on the templates prepared above. Touch a sterile toothpick to a clone from round 2 and then to the same position on the grid for each of the two array plates. Pick at least 10 clones from each plate in round 2, deliberately choosing some negative clones as well as positive ones.
3. Incubate the plates ON at 37°C. Then store one at 4°C as a source of phage and blot the other. After screening, transfer each entire positive placque from the stored plate to 250 µL SM, and add 10 µL of $CHCl_3$.

3.3. Isolation of Phage DNA (Craik et al., 1984)

1. Centrifuge a stationary phase culture (5 h–ON) of the host cells for 10 min at $1000 \times g$. Resuspend the pellet in an equal vol of SM buffer.

Isolation and Structure of Genes

2. Mix 100 µL of cells with 50 µL of the phage isolated from the array described above (about 1×10^6 pfu), and incubate for 20 min at 37°C. Add this cell suspension to 40 mL of 37°C media, and incubate at 37°C with vigorous shaking ON. During this time, the culture should become turbid and then clear somewhat, and small "strings" should appear at the bottom of the flask. Add 100 µL of $CHCl_3$ to complete the lysis of the host cells, shake for 15 min at 37°C, and centrifuge 10 min at $1000 \times g$ to remove the cell debris.
3. Combine the supernatant with 40 µL DNase I (1 mg/mL) in 0.15M NaCl, 50% glycerol, and with 4 µL of 10 mg/mL DNase-free RNase, and incubate for 30 min at 37°C.
4. Add 5 mL 5M NaCl and 5 mL 50% polyethylene glycol 8000, and incubate for 2 h on ice to precipitate the phage.
5. Centrifuge the samples for 10 min at $1000 \times g$, pour off the supernatant, and wipe the walls of the tube with a Kimwipe to remove the excess polyethylene glycol solution.
6. Resuspend the pellets in 400 µL proteinase K buffer: 10 mM Tris pH 7.8, 5 mM EDTA, 0.5% SDS. Add 2 µL of 10 mg/mL proteinase K in water, and incubate for 20 min at 37°C.
7. Extract the proteinase K, digest twice with an equal volume of $CHCl_3$: UNC-phenol, 1:1, and once with an equal volume of $CHCl_3$.
8. Layer 1 mL of room temperature ethanol on the solution, and spool the DNA precipitating at the interface by gently stirring the solution with a Pasteur pipet that has been heated until it has sealed and formed a little bulb. Invert the pipet to allow any excess ethanol to run off, but do not allow the DNA to dry. Dissolve the spooled DNA in 100–200 µL of 10 mM Tris pH 7.6. Run 1 µL aliquots on an agarose gel along with 0.5 µg aliquots of λ DNA to estimate the yield. Normally, we can produce about 100–200 µg of phage DNA with this procedure.

4. Structure Determination of Genomic Clones

4.1. Preliminary Mapping

In a preliminary screen to determine which clones are alike, take 2 μg aliquots of DNA from each, and digest them with one or two restriction enzymes with six-base restriction sites, chosen from those enzymes that do not have sites within the arms of the vector phage. Reasonable enzymes would be those in the polylinker region, *Eco* RI, *Bam* HI, or *Sal* I. Different restriction enzymes have many different buffer requirements, the most important of which is the salt concentration. For convenience in mapping studies using enzyme combinations, we first attempt to use one of three standard buffers: *Hae* III buffer (6 mM Tris pH 7.6, 6 mM MgC1$_2$, 50 mM NaC1, 6mM β-mercaptoethanol), 2× *Hae* III buffer, or 3× *Hae* III buffer. In general, we use the version of *Hae* III buffer that has a salt concentration closest to that recommended by the manufacturer for the enzyme in use. For enzymes that demand widely different salt concentrations, we digest first for 2 h with the low-salt enzyme, and then increase the salt and digest for an additional 2 h with the higher salt one.

Following digestion, the samples are diluted with 6 μL of loading buffer, and together with molecular weight standards, applied to a 1.2% agarose—Howley's buffer gel. The buffer is recirculated if possible. After 16 h of electrophoresis at 50 volts, the gel is stained for 10 min with dilute ethidium bromide solution, and the bands are visualized with ultraviolet light. Electrophoresis is then continued until a 1 kb DNA fragment has moved approximately 10 cm. The mobility of the mol wt standards is then plotted against the log of the number of bases in each. The mol wt of the unknown phage fragments can be calculated from the curve thus generated.

It is quite important that the cloned DNA be completely digested, since partial digestions can lead to serious problems in interpreting mapping experiments. A complete restriction digest is characterized on an ethidium bromide stained gel by an array of bands whose intensity decreases regularly with their mol wt. A partial digest is indicated when bands are observed flanked on either side by more intense bands. When the problem can be traced to impurities in the DNA preparation, which inhibit the digestion, it is sometimes possible to reduce the problem by diluting the DNA two- to fourfold, keeping the enzyme concentration constant.

4.2. Southern Blotting (Southern, 1975)

When several different clones have been identified for analysis, digest them as above and separate the fragments by gel electrophoresis. The gel is prepared for blotting according to the manufacturer's instructions for the membrane that is to be used. For nitrocellulose, the gel is gently agitated in $0.2M$ NaOH, $0.6M$ NaCl for 45 min to denature the DNA. The gel is rinsed in water and immersed in $1M$ Tris pH 7.6, $0.6M$ NaCl for 45 min, or until a strip of pH paper touched to the surface registers approximately pH 8. The gel is then trimmed, cutting off material above the top of the wells, along the bottom and sides, and in any unused lanes.

During the treatment of the gel, a large glass baking dish is assembled with a 20-cm wide glass plate that can be suspended across its top, and four pieces of 3-mm thick filter paper are cut. One, the wick, should be somewhat wider than the length of the gel and should be long enough to drape over the glass plate to the baking dish bottom. The second should be 2–3 cm larger than the trimmed gel and the last two should be 1–2 cm larger than the gel. A piece of nitrocellulose is also cut 1–2 cm larger than the gel. The nitrocellulose is then wet by capillary action, as in section 3.2.1.

To assemble the blot, the baking dish is partially filled with $20\times$ SSC; the wick is immersed in the buffer and draped over the glass plate. A pipet is rolled over the wick to remove any air bubbles trapped beneath it. Filter #2 is soaked with $20\times$ SSC and centered on the gel; the gel holder is inverted and the paper-backed gel is positioned on the wick. The gel is "rolled with a pipet" to remove any air bubbles, and is surrounded with pieces of plastic wrap to prevent any overlapping portions of nitrocellulose from touching the wick and short-circuiting the blotting process. Then the nitrocellulose is carefully lowered onto the gel. Once the membrane touches the gel surface, it should not be moved. After any air bubbles have been removed, one of the last two sheets of filter paper is placed on the membrane and gently rolled with a pipet until it is completely wet. This process is repeated with the last sheet of filter paper. Then the sandwich is topped with a stack of paper towels, a glass plate, and a weight such as a bottle containing a liter of liquid.

After 15–18 h, the paper towels are removed, the sandwich is inverted, and the exact position of the gel on the nitrocellulose is marked using a ball-point pen. A sheet of standard size nitrocellulose (20× 18 cm) is somewhat unwieldly to handle, so we routinely cut it into four pieces at this point, marking the pieces and making the horizontal cut at a slight angle so that the pieces can be put back together only one way. The membrane is air-dried and baked in a vacuum oven at 80°C for 2 h. It is hybridized, washed, and exposed to film as described in section 3.2.1.

4.3. Building a Complete Map

Examine the autoradiogram as well as the picture of the gel, and choose a few clones for more extensive mapping studies. Be certain that the Southern blot of the clones is consistent with a Southern blot of 10 µg uncloned genomic DNA. Inconsistencies may suggest rearrangement of the DNA during the cloning protocol. After selecting clones for further analysis, choose a series of six-base restriction enzymes, for example, *Eco* RI, *Bam* HI, *Hind* III. Digest each clone with each enzyme alone and with all combinations of two enzymes. Since the apparent mol wt of a fragment varies by about 10% between gels, it is important to place all of the digests for a single clone on the same agarose gel. One can then build a map by comparing fragment sizes. In the simplest case, if a probe reacts with a 5 kb *Eco* R1 fragment, with a 4 kb *Bam* HI fragment and, in a mixed digest of the two, with a 1 kb *Eco* R1-*Bam* H1 fragment, then the beginning of the map of the insert must be as shown in Fig. 2.

4.4. Subcloning Segments of the Gene

Once a detailed restriction map of the isolated DNA is available, it is possible to identify appropriate fragments to subclone for sequence analysis. Ideally, subcloning into the single-stranded sequencing vector M13 is done with relatively small genomic DNA fragments, although one is rarely so fortunate as to be able to subclone from λ directly into M13. The maximum insert that is reported to be stable in M13 is approximately 1.0 kb (Maniatis et al, 1982), although we have often obtained M13 clones with inserts up to 2.6 kb.

Generally, the best approach for obtaining well-defined DNA fragments for M13 subcloning is to subclone larger fragments into a

Isolation and Structure of Genes

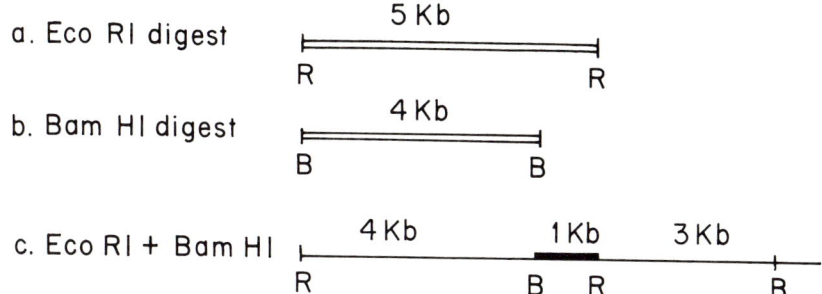

Fig. 2. Map of a genomic DNA region, which contains the following fragments that react with a cDNA probe: (a) a 5 kb Eco R1 fragment, (b) a kb Bam H1 fragment, (c) a 1 kb Eco R1-Bam H1 fragment.

plasmid. After detailed characterization, these subclones are further subcloned into M13 for sequencing. For the initial subcloning, several features of the vector are quite important. The size of the vector is relevant, i.e., try to select a vector with a significantly different size from the insert to be subcloned. The most commonly used vectors have been pBR322 and pUC18/19, but some other useful plasmid vectors are pUC118/119, pBluescript (Stratagene), or pGEM (Promega), since these vectors can be used as double-stranded DNA molecules for restriction mapping, or as double- or single-stranded molecules for DNA sequencing. The multi-linker region in the pUC series, pBluescript, and pGEM is extremely useful, since these plasmids have been engineered to contain a set of unique restriction enzyme sites in the multi-linker region. Thus, using a relatively large set of restriction enzymes, inserts can be generated and ligated into a defined area of the plasmid, and excised using the same enzyme(s) or flanking enzyme sites within the multi-linker region.

In many cases, restriction enzymes generate DNA fragments with sticky ends, which ligate quite easily. These can generally be recleaved with the same enzyme, and the insert excised. When a restriction enzyme that generates blunt ends is used, ligation is more difficult, and the site in the multiple-cloning region for that enzyme is often destroyed by the ligation. Thus, to excise the insert, another enzyme in the multiple-cloning region must be utilized, preferably one that does not have sites within the insert itself.

Subcloning can be achieved either by "shotgun" cloning or by purifying the insert to be subcloned. For complete mapping of a λ insert, subcloning all fragments from the large insert is important. In this case, or when it appears likely that it would be difficult to obtain a purified insert for subcloning, "shotgun" cloning is preferable. The clone is digested with the appropriate enzyme(s), and the digest is added directly to plasmid vector, ligated, and used for bacterial transformation. A large number of clones should be purified and grown up, in order to isolate at least one subclone containing each of the insert fragments. Plasmid DNA from these subclones can be prepared (see Chapter 3) and analyzed for insert size by digestion with the appropriate restriction enzymes and electrophoresis on agarose gels. Exon-containing subclones can be identified by Southern blot analysis. It may be desirable to subclone a purified insert if only a single, specific subclone is desired. However, subcloning an isolated insert may be less successful than shotgun subcloning, for example if there are contaminants in the purified insert DNA that inhibit ligation or some other step in the subcloning. In either subcloning procedure, it is valuable to use a plasmid that allows a blue/white selection for plasmids with inserts (e.g., pUC), but it is particularly useful when subcloning isolated inserts, since all clones with inserts should be the desired clones.

A generalized protocol for "shotgun" subcloning a λ genomic clone into pUC is presented here. This is adapted from protocols provided by Maniatis et al. (1982) and Davis et al. (1986).

1. Digest 2 µg of the λ cloned DNA with 10–12 U of the desired enzyme(s) in a final reaction vol of 20 µL for 2 h at 37°C.
2. Digest 1 µg pUC18 with 5–10 U of the appropriate enzyme in 10 µL final vol.
3. Run a test gel to assess whether all digestions were complete. Electrophorese 1 µL aliquots of the vector DNA and the λ clone DNA digests on a 1% agarose gel in TBE for 2–16 h, with appropriate mol wt standards.
4. If digestions appear to be complete, dephosphorylate the vector to prevent its religation without inserts. (This step is not usually necessary when ligating fragments generated by two restriction enzymes, since the vector can religate only in the presence of the insert or the small piece of the multiple-cloning

region that was excised.) Add to the vector digest mix 2 µL 10× alkaline phosphatase buffer (10×: 0.5M Tris pH 9, 10 mM MgCl$_2$, 1 mM ZnCl$_2$), 1 µL calf intestinal alkaline phosphatase (0.3 U/µL) and 7 µL water. Mix and incubate for 30 min at 37°C.
5. Bring vol of the vector and insert reactions to 50 µL with water, and extract once with UNC-phenol/chloroform. Add 25 µL 7.5M ammonium acetate, and precipitate the DNA with 2 vol ethanol at –20°C.
6. Resuspend vector DNA, and insert DNA each in 20 µL TE.
7. Add 20 µL digested λ DNA, 2 µL digested vector, 3 µL 10× ligase buffer, and 1.0 U T4 DNA ligase in 30 µL final volume. Incubate ON at 16°C. This is approximately a 1:1 *molar* ratio of insert:vector. When ligating preparations of isolated insert, rather than shotgunned λ DNA, mix insert and vector DNA at an insert:vector ratio of 3:1. When ligating DNA fragments with blunt ends, significantly higher amounts of ligase will be needed, and the insert:vector ratio may need to be increased.

Appropriate controls to assess the efficiency of the subcloning
1. no ligase
2. no phosphatase
3. dephosphorylated vector ligated with no insert

To prepare competent cells, grow bacteria in LB broth to an O.D.$_{600}$ of 0.3–0.5. Centrifuge cells, and resuspend pellet in ice-cold 50 mM CaCl$_2$ one-half that vol for 30 min on ice. Centrifuge cells, and resuspend in one-tenth the original volume of 50 mM CaCl$_2$; cells can be stored for 1–4 d at 4°C. To transform cells, incubate 0.1 µg of ligated plasmid vector DNA with 200 µl of competent cells. Incubate on ice for 30–60 min. Heat cells in a 42°C waterbath for 2 min, add 1 mL LB broth, and grow cells at 37°C for 45 min with shaking to allow expression of the antibiotic resistance gene. Grow cells on plates containing the appropriate antibiotic, incubate ON at 37°C, and isolate single colonies for further analysis. Plasmids of interest can be maintained for long periods at –70°C in transformed cell stocks containing 20% sterile glycerol.

4.5. Sequencing Subcloned Genomic DNA

Much of the DNA sequencing by these two laboratories has been by dideoxy DNA sequencing of M13 clones. This protocol has been described extensively in Chapter 3, and will not be repeated here. Two other DNA sequencing protocols are used when necessary. The chemical cleavage protocol of Maxam and Gilbert (1980) is valuable when certain DNA fragments are refractory to dideoxy DNA sequencing.

Another DNA sequencing protocol of value is double-stranded DNA sequencing (Chen and Seeburg, 1985). The pBluescript, pGEM, and pUC118/119 vectors can be prepared as single- or double-stranded DNA molecules. The single-stranded DNA preparation protocol has a significantly lower yield than the double-stranded DNA protocol, and in some instances, using the double-stranded DNA template is preferable. A negative of this protocol can be the greater difficulty in reading certain DNA sequences. Nevertheless, the double-stranded sequencing protocol is extremely useful when subcloning to demonstrate the presence of an insert and the orientation of the insert in the vector. We have used it primarily with avian myeloblastosis virus (AMV) reverse transcriptase. When extensive sequence information is needed, utilization of both AMV reverse transcriptase and Klenow in parallel incubations may be helpful (Mierendorf and Pfeffer, 1987).

Sequencing gels are run essentially identically to those described in Chapter 3, except that for many samples, buffer gradient gels are utilized (Biggin et al., 1983, as adapted by BioRad). This is a gradient of $0.5\times$ TBE to $1.5\times$ TBE with 6% acrylamide and $7.67M$ urea. The light solution (75 mL/38×50 cm gel) contains $0.5\times$ TBE, 6% acrylamide, $7.67M$ urea. The heavy solution (20 mL/gel) contains $2.5\times$ TBE, 6% acrylamide, $7.67M$ urea and 10% sucrose. Just before pouring, 0.25% ammonium persulfate and 1 µL TEMED are added to each solution. 12 mL of each solution are mixed and poured into the gel plates, and then 50 mL of the light solution are poured into the apparatus to produce a gradient. We have been able to resolve as many as 450 nucleotides on a 50-cm gel with this protocol, because the short DNA fragments are compressed, relative to nongradient gels.

5. Investigation of Transcriptional and Regulatory Elements of Genomic Clones

In order to identify the transcription initiation site of exon 1, the end of the mRNA can be mapped using either S1 nuclease (Berk and Sharp, 1977) or primer extension (Calzone et al, 1987). We have used S1 nuclease mapping more extensively and will discuss it below. Since S1 nuclease digests single-stranded nucleic acids, it is possible to hybridize RNA with radioactive single-stranded genomic fragments containing exon one and flanking DNA sequences. The reaction mixture is then digested with S1 nuclease to digest any unhybridized probe. By knowing the position of the 3'-end of the exon 1 probe and the size of the protected hybridized probe, it is possible to establish the 5'-end of the mRNA. When using primer extension, an oligonucleotide synthesized to hybridize with a known segment of the mRNA or a single-stranded genomic (or cDNA) fragment that ends at a known site within the mRNA can be hybridized with the mRNA and used as a primer for reverse transcription. The primer extension will terminate at the end of the mRNA, and the size of the extended primer will identify that nucleotide. Both S1 nuclease and primer extension should provide identical information, but it is helpful to use both approaches, since ambiguities can arise. For example, S1 nuclease mapping of the end of exon 1 of the myelin PLP gene indicated that the major transcription initiation site (A) produced a 164-nucleotide exon 1 segment, with several weaker initiation sites 30 (B) and 65 nucleotides (C) downstream (Macklin et al., 1987). The same three initiation sites were observed by primer extension, but the major initiation site appeared to be site B (Milner et al., 1985).

The S1 nuclease mapping protocol used is an adaptation of the method of Berk and Sharp (1977) using probe generated by the procedure of Ley et al. (1982). Uniformly labeled probe is generated from insert DNA in M13 by a standard sequencing protocol (*see* Chapter 3), using 50 µCi α-^{32}P dATP, 0.05 mM dATP, 0.5 mM dCTP, 0.5 mM dGTP, and 0.5 mM dTTP, and no dideoxynucleotides. The optimal incubation time should be determined for each probe, based on its length. Single-stranded probe of a defined length is needed, and it can be generated by digesting the newly synthesized molecule with an appropriate restriction enzyme. Generally, the restriction enzyme is selected from the multi-linker region on the side of the insert opposite to that of the sequencing primer, and it will produce a single cut in the M13 molecule. The digested sample is electrophoresed on a 4% acrylamide-50% urea-TBE gel, which will separate the nonradioactive, single-stranded M13 template from the shorter radioactive probe. A gel piece containing the probe is excised, and DNA is eluted. Sodium acetate (0.2M) is added, and the probe is UNC-phenol/chloroform extracted and precipitated with 2 vol of ethanol. Purified probe is resuspended in 80% formamide, 40 mM PIPES, 1 mM EDTA with either 20–50 μg total RNA, or 2–5 μg poly A^+ RNA with approximately 20 μg carrier tRNA. NaCl is added to a final concentration of 0.4M, and the sample is incubated at 75°C for 15 min, and immediately transferred to a 52–54°C water bath ON for hybridization. For S1 nuclease digestion, 300 μL S1 nuclease buffer is added (280 mM NaCl, 50 mM Na Acetate, 4.5 mM $ZnSO_4$, 20 μg/mL salmon sperm DNA) containing 175 U S1 nuclease (500 U/mL, final concentration). Samples are incubated at 37°C for 60 min. The protected double-stranded material is UNC-phenol/chloroform extracted and precipitated with ethanol, and then analyzed on TBE gels containing 4, 6, or 8% acrylamide and 50% urea with appropriate radiolabeled mol wt standards. After electrophoresis, the gel is fixed, dried, and autoradiographed. The specific conditions described here were optimized for myelin PLP mRNA. For each RNA study, initial characterizations of the hybridization time, temperature, and formamide concentration must be conducted, and the optimal S1 nuclease digestion time must be determined. Essential controls include hybridized samples with no S1 nuclease digestion, and probe digested in the absence of mRNA.

In addition to mapping and sequencing the exons of a particular gene and determining the transcription initiation site, a major focus for gene expression studies includes an analysis of the regulatory elements. A number of consensus sequences in the upstream region of the gene have been identified. The most common of these sequences are two promoter elements, the TATAA element, an AT-rich short sequence, which is found 25–30 bases upstream from the transcription initiation site (Corden et al., 1980), and the CAAT element (consensus sequence, CCAAT, Efstratiadis et al., 1980), which is often found 80 bases upstream of the

Isolation and Structure of Genes

transcription initiation site. The CAAT element is found less often than the TATAA element, and its precise function is unknown. The position of these two DNA sequences, relative to the transcription initiation site, is invariant. In contrast, enhancer sequences have been found in widely varying positions, relative to the transcription initiation site, both upstream and downstream of the gene and within intron regions.

Because of their invariant positions, once the transcription initiation site is established, sequencing 100 nucleotides upstream of exon 1 should identify the TATAA and CAAT elements. It may also be possible to identify other short nucleotide sequences that are apparent binding sites for gene regulators such as glucocorticoids. However, in order to map other regulatory sequences of a gene, other approaches are needed. If there is a suitable cultured cell system that can be used, the regulatory elements can be mapped by transfecting DNA constructs containing sequences from the gene linked to a reporter gene. DNA uptake can be induced by calcium phosphate coprecipitation, DEAE-dextran treatment, electroporation, or viral infection. One frequently used reporter gene is the bacterial gene chloramphenicol acetyltransferase (CAT) (Gorman et al., 1982). CAT expression can be measured by enzyme assay in cells after DNA uptake. An advantage of the enzyme assay is its sensitivity, although since it is a measure of enzyme activity, it is only an indirect measure of transcription. Once the CAT assay has been established in a particular cell system, one can delete or add segments of genomic DNA to the genomic DNA/CAT gene construct, and then measure changes in CAT levels. Other reporter genes are available, and other approaches such as generation of transgenic mice can be used to characterize regulatory elements of genes. A more extensive description of these protocols is provided by Rosenthal (1987).

Acknowledgments

The authors would like to thank J. Norman Hanson for guidance in the utilization of the disulfide acrylamide gel protocol. This work was supported, in part, by NIH grants NS23715, NS25304, a career development award to WBM (NS1089), and National Multiple Sclerosis Society grant RG1910.

References

Berger S. L. and Kimmel A. R. (1987) *Methods in Enzymology: Guide to Molecular Cloning Techniques,* **Vol. 152.** Academic, San Diego.

Berk A. J. and Sharp P. A. (1977) Sizing and mapping of early adenovirus mRNAs by gel electrophoresis of S1 endonuclease-digested hybrids. *Cell* **12,** 721–732.

Biggin M. D., Gibson T. J., and Hong G. F. (1983) Buffer gradient gels and ^{35}S label as an aid to rapid DNA sequence determination. *Proc. Natl. Acad. Sci. USA* **80,** 3963–3965.

Blattner F. R., Williams B. G., Blechl A. E., Denniston-Thompson K., Faber H. E., Furlong L.-A., Grunwald D. J., Diefer D. O., Moore D. D., Sheldon E. L., and Smithies O. (1977) Charon phages: Safer derivatives of bacteriophage lambda for DNA cloning. *Science* **196,** 161–169.

Burke D. T., Carle G. F., and Olson M. V. (1987) Cloning of large segments of exogenous DNA into yeast by means of artificial chromosome vectors. *Science* **236,** 806–812.

Calzone F. J., Britten R. J., and Davidson E. H. (1987) Mapping of gene transcripts by nuclease protection assays and cDNA primer extension. *Methods Enzymol.* **152,** 611–632.

Campagnoni C. W., Carey G. D., and Campagnoni A. T. (1978) Synthesis of myelin basic proteins in the developing mouse brain. *Arch. Biochem. Biophys.* **190,** 118–125.

Carson J. H., Nielson M. L., and Barbarese E. (1983) Developmental regulation of myelin basic protein expression in mouse brain. *Develop. Biol.* **96,** 485–492.

Chen E. Y. and Seeburg P. H. (1985) Supercoil sequencing: A fast and simple method for sequencing plasmic DNA. *DNA* **4,** 165–170.

Clarke L. and Carbon J. (1976) A colony bank containing synthetic Col E1 hybrid plasmids representative of the entire E coli genome. *Cell* **9,** 91–99.

Cochet M., Gannon F., Hen R., Maroteaux L., Perrin F., and Chambon P. (1979) Organisation and sequence studies of the 17-piece chicken conalbumin gene. *Nature* (London) **282,** 567–574.

Collins F., and Hohn B. (1978) Cosmids: A type of plasmid gene-cloning vector that is packageable in vitro in bacteriophage lambda heads. *Proc. Natl. Acad. Sci.* **75,** 4242–4246.

Cooke N. E. and Baxter J. D. (1982) Structured analysis of the prolactin gene suggest a separate origin for its 5' end. *Nature* **297,** 603–606.

Corden J., Wasylyk B., Buchwalder A., Sassone-Corsi P., Kedinger C., and Chambon P. (1980) Promoter sequences of eukaryotic protein-coding genes. *Science* **209**, 1406–1414.

Craik C. S., Choo Q.-L., Swift G. H., Quinto C., MacDonald R. J., and Rutter W. J. (1984) Structure of two related rat pancreatic trypsin genes. *J. Biol. Chem.* **259**, 14255–14264.

Davis L. G., Dibner M. D., and Battey J. F. (1986) *Basic Methods in Molecular Biology*. Elsevier, New York, pp. 11–114.

DiLella A. G. and Woo S. L. C. (1987) Cloning large segments of genomic DNA using cosmid vectors. *Methods Enzymol.* **152**, 199–212.

Dretzen G., Bellard M., Sassone-Corsi P., and Chambon P. (1981) A reliable method for the recovery of DNA fragments from agarose and acrylamide gels. *Anal. Biochem.* **112**, 295–298.

Efstratiadis A., Posakony J. W., Maniatis T., Lawn R. M., O'Connell C., Spritz R. A., DeRiel J. K., Forget B. G., Weissman S. M., Slightom J. L., Blechl A. E., Smithies O., Baralle F. E., Shoulders C. C. and Proudfoot N. J. (1980) The structure and evolution of the human β-globin gene family. *Cell* **21**, 653–668.

Feinberg A. P. and Vogelstein B. (1983) A technique for radiolabeling DNA restriction endonuclease fragments to high specific activity. *Anal. Biochem.* **132**, 6–13.

Frischauf A. M., Lehrach H., Poostka A., and Murray N. (1983) Lambda replacement vectors carrying polylinker sequences. *J. Mol. Biol.* **170**, 827–842.

Frischauf A. M., Murray N., and Lehrach H. (1987) Lambda phage vectors—EMBL series. *Methods Enzymol.* **153**, 103–115.

Gilliam T. C., Tanzi R. E., Haines J. L., Bonner T. I., Faryniarz A. G., Hobbs W. J., MacDonald M. E., Cheng S. V., Folstein S. E., Conneally P. M., Wexler N. S., and Gusella J. F. (1987) Localization of the Huntington's disease gene to a small segment of chromosome 4 flanked by D4S10 and the telomere. *Cell.* **50**, 565–571.

Gorman C., Moffat L., and Howard B. (1982) Recombinant genomes which express chloramphenicol acetyl transferase in mammalian cells. *Mol. Cell. Biol.* **2**, 1044–1051.

Hansen J. N. (1981) Use of solubilizable acrylamide disulfide gels for isolation of DNA fragments suitable for sequence analysis. *Anal. Biochem.* **116**, 146–151.

Hoffman E. P., Brown R. H., and Kunkel L. M. (1987) Dystrophin: The protein product of the Duchenne muscular dystrophy gene in mice and humans. *Science* **238**, 347–350.

Ish-Horowitz D. and Burke J. F. (1981) Rapid and efficient cosmid vector cloning. *Nuc. Acids. Res.* **9**, 2989–2998.

Karin M., Haslinger A., Holtgreve H., Richards R. I., Krauter P., Westphal H. M. and Beato M (1984) Characterization of DNA sequences through which cadmium and glucocorticoid hormones induce human metallothionein-II$_A$ gene. *Nature* **308**, 513–519.

Koenig M., Hoffman E. P., Bartelson C. J., Monaco A. P., Feener C., and Kunkel L. M. (1987) Complete cloning of the Duchenne muscular dystrophy (DMD) cDNA and preliminary genomic organization of the DMD gene in normal and affected individuals. *Cell* **50**, 509–517.

Lau Y. F. and Kan Y. W. (1983) Versatile cosmid vectors for the isolation, expression and rescue of gene sequences. *Proc. Natl. Acad. Sci. USA* **80**, 5225–5229.

Leach D. R. F. and Stahl F. W. (1983) Viability of lambda phages carrying a perfect palindrome in the absence of recombination nucleases. *Nature* (London) **305**, 448–451.

Leff S. E., Evans R. M., and Rosenfeld M. G. (1987) Splice commitment dictates neuron-specific alternative RNA processing in calcitonin/CGRP gene expression. *Cell* **48**, 517–524.

Ley T. J., Anagnou N. P., Pepe G., and Nienhuis A. W. (1982) RNA processing errors in patient with beta-thalassemia. *Proc. Natl. Acad. Sci. USA* **79**, 4775–4779.

Macklin W. B., Campagnoni C. W., Deininger P. L., and Gardinier M. V. (1987) Structure and expression of the mouse myelin proteolipid protein gene. *J. Neurosci. Res.* **18**, 383–394.

Maniatis T., Fritsch E. F., and Sambrook J. (1982) *Molecular Cloning, a Laboratory Manual*, Cold Spring Harbor Laboratory, Cold Spring Harbor, New York.

Martin J. B. and Gusella J. F. (1986) Huntington's disease: Pathogenesis and management. *New England J. Med.* **315**, 1267–1276.

Maxam A. M. and Gilbert W. (1980) Sequencing end-labeled DNA with base-specific chemical cleavages. *Methods Enzymol.* **65**, 499–560.

Mierendorf R. C. and Pfeffer D. (1987) Direct sequencing of denatured plasmid DNA. *Methods Enzymol.* **152**, 556–562.

Milner, R. J., Lai, C., Nave, K.-A., Lenoir, D., Ogata, J., and Sutcliffe, J. G. (1985) Nucleotide sequences of two mRNAs for rat brain myelin proteolipid protein. *Cell* **42**, 931–939.

Niall, H. D., Hogan, M. L., Sauer, R., Rosenbaum, I. Y., and Greenwood, F. C., (1971) Sequences of pituitary and placental lactogenic and growth hormones: Evolution from a primordial peptide by gene duplication. *Proc. Natl. Acad. Sci.* **68**, 866–868.

Park, I., Schaeffer, E., Sidoli, A., Baralle, F. E., Cohen, G. N., and Zakin, M. M. (1985) Organization of the human transferrin gene: Direct evidence that it originated by gene duplication. *Proc. Natl. Acad. Sci.* **82,** 3149–3153.

Rigby, P.W.J., Dieckmann, M., Rhodes, C., and Berg, P. (1977) Labeling deoxyribonucleic acid to high specific activity in vitro by nick translation with DNA polymerase I. *J. Mol. Biol.* **113,** 237–251.

Rosenthal, N. (1987) Identification of regulatory elements of cloned genes with functional assays. *Methods Enzymol.* **152,** 704–720.

Smith, H. O. (1980) Recovery of DNA from gels. *Methods Enzymol.* **65,** 371–381.

Southern, E. M. (1975) Detection of specific sequences among DNA fragments separated by gel electrophoresis. *J. Mol. Biol.* **98,** 503–517.

Stahl, F. W., Craseman, J. M., and Stahl, N. N. (1975) Rec-mediated recombinational hot spot activity in bacteriophage lambda. *J. Mol. Biol.* **94,** 203–212.

Stuart, G. W., Searle, P. F., and Palmiter, R. D. (1985) Identification of multiple metal regulatory elements in mouse metallothionein-I promoter by assaying synthetic sequences. *Nature* **317,** 828–831.

Takahash, N., Roach, A., Teploco, D. B., Prusiner, S. B., and Hood, L. (1985) Cloning and characterization of the myelin basic protein gene from mouse: One gene can encode both 14kd and 185kd MBPs by alternate use of exons *Cell* **42,** 139–148.

Webster, C., Silberstein, L., Hays, A. P., and Blau, H. M. (1988) Fast muscle fibers are preferentially affected in Dughenne muscular dystrophy. *Cell* **52,** 503–513.

Wieslander, L. (1979) A simple method to recover high molecular weight RNA and DNA after electrophoretic separation in low gelling temperature agarose gels. *Anal. Biochem.* **98,** 305–309.

Winberg G. and Hammarskjold M.-L. (1980) Isolation of DNA from agarose gels using DEAE-paper. Application to restriction site mapping of adenovirus type 16 DNA. *Nucl. Acids Res.* **8,** 253–264.

Wu R., Jay E., and Roychoudhury R. (1976) Nucleotide sequence analysis of DNA. *Methods in Cancer Res.* **12,** 88–171.

Wyman A. R., Wertman K. F., Barker D., Helms C., and Petri W. H. (1986) Factors which equalize the representation of genome segments in recombinant libraries. *Gene* **49,** 263–271.

Wyman A. R. and Wertman K. F. (1987) Host strains that alleviate underrepresentation of specific sequences: overview. *Methods Enzymol.* **152,** 173–180.

Gene-Mapping Techniques

Robert S. Sparkes

1. Introduction

The focus of this chapter is on human gene mapping, but in principle, the information applies to mapping in other mammals, including other primates and rodents. Gene mapping is the discipline that assigns genes to chromosomes and parts of chromosomes. There has been recent rapid progress in mapping human genes, and it now appears technically possible to assign any gene to a chromosome and to a small region of a given chromosome. This has become possible largely because of the ability to isolate and manipulate specific DNA sequences.

Efforts have been under way for a number of years to periodically update and evaluate gene mapping information. The major attempt at this has been at the International Workshops on Human Gene Mapping, the latest of which was held in Paris (Human Gene Mapping 9, published in *Cyto-genetics and Cell Genetics*, 1987). A computer record of these meetings and an ongoing record of new mappings is kept by the Howard Hughes Medical Institute, New Haven Gene Mapping Library (25 Science Park, New Haven, CT 06511, Tel. 203-786-5515). Victor McKusick, MD (The Johns Hopkins University School of Medicine) maintains his own frequently updated newsletter, which he distributes to workers in the field.

Gene-mapping information is important at both the clinical and basic biological levels. In the clinical sphere, gene-mapping information can be used to detect genetic heterogeneity (genocopies), which may have important implications for diagnosis and genetic counseling. Genocopies refer to those situations that appear to have the same phenotype, but are the result of mutations at different loci. Gene-mapping information may also be helpful to identify persons at risk for a genetic disease in the presymptomatic stage, which is usually accomplished through linkage analysis.

Linkage analysis may also be used for prenatal diagnosis of a genetic trait that is not expressed at an early stage of development or is in a tissue unavailable for direct testing. Gene-mapping information is useful for carrier detection in recessive conditions in which direct expression of the affected gene cannot be detected. Mapping information can be used to identify the parent from whom a chromosomally abnormal child originated; this has implications for eventual recognition of factors causing chromosome abnormalities with future prevention of other affected children. Insight may also be achieved into the understanding of pathogenesis of the abnormal phenotype that is associated with chromosome abnormalities. For example, is the abnormal phenotype the result of imbalance of specific genes on the affected chromosome?

There are also many reasons why gene-mapping information is important for basic biological considerations. It can increase knowledge of factors related to genetic recombination with implications for evolution. Factors that may affect genetic crossing over can be identified (these may include sex, age, race, heterochromatin, and centromere effects). Gene-map distances can be evaluated through frequency of crossing over or actual physical distances as represented by the number of DNA bases that separate genetic loci; there is already some information that these two may not be the same or equal for different chromosomes and/or genetic loci. Gene-map information will also help to evaluate whether a gene is affected by its location, i.e., position effect. Chromosome structure appears to be complex and relatively conserved through evolution. Understanding the position of genes in the various chromosomes, such as their relation to banding patterns, will give improved understanding of chromosome structure and function. It will also help to determine whether gene locations are related in some specific way to their function. Identification of genetic control and regulatory functions for specific genes can be evaluated and clarified. It is becoming clear that chromosome rearrangements associated with at least some forms of cancer are related to specific genes on a chromosome. Thus, knowing which genes are involved may give insight into pathogenesis, as has been possible with chronic myelogenous leukemia, and gene-map information may help in identifying and isolating these genes. The role of pseudogenes may be clarified, and the effect of heterochromatin on gene expression may be better understood.

2. Technical Advances

As so often is the case in science, recent technical developments have made it increasingly possible and easy to map human genes. Mapping through genetic linkage analysis has been greatly facilitated by the development of improved statistical analyses combined with more powerful computer capabilities. The ability to store viable cells in cell cultures in liquid nitrogen has made disease studies easier. The development of various chromosome banding techniques has made it possible to identify each and every chromosome, and also to identify specific regions and bands for regional mapping of genes. The use of somatic-cell hybridization techniques has been a mainstay in assigning human genes to chromosomes. Finally, there has been the development of new phenotypic and especially DNA polymorphisms, which has revitalized human genetic linkage studies.

3. Anatomy of the Genome

The gene map represents the anatomy of the genome, and as such, it is important that it be as complete and accurate as possible. The map can be considered at different levels. At the most gross level, one can distinguish whether a gene is located on the X chromosome or on the autosomes. A slightly finer mapping will permit assignment of a gene to a specific autosome and, furthermore, to part of an autosome. One may also understand the map at the molecular level with references to the sequences of a gene and its contiguous segments. Finally, there will be the full sequence of the total human genome; because of variability in DNA sequences, it can be anticipated that the difference between sequences in different individuals will be great, wherein lies part of the controversy in regard to the complete sequencing of the human genome.

The size of the human genome can be considered in terms of the physical map or the total number of base pairs, or in terms of meiotic recombination, which is generally indicated in terms of centimorgans. The genome can also be reflected by the total of functional genes. It has been estimated that the human haploid genome contains about 3.3 billion base pairs. It has been estimated that there are between 50,000 and 100,000 functional genes in the human genome. In relation to the number of centimorgans present in the human genome, it has been estimated that this may be approximately 3,300 on average, but that this probably differs between males and females. The number of centimorgans on each chromosome will also vary in relation to its size, with the larger ones having a greater number. Estimates have also been made as to how many centimorgans may be present in a chromosome band, and this will depend upon the number of bands that can be demonstrated in the karyotype; when 600 bands are seen, each band would then represent about five centimorgans and probably contain between 100 and 200 genes. One can also calculate that each centimorgan on average will represent approximately 1,000,000 base pairs and, therefore, each chromosome band in a 600-band preparation will represent approximately 5,000,000 base pairs.

With the rapid expansion of knowledge regarding genes and their identification, there has been a pressing need to develop a consistent nomenclature for gene identification. Probably the best exposition of the basis for the accepted nomenclature is found in the report of the Human Gene Mapping 8 (*Cytogenetics and Cell Genetics*, 1985). Genes with known function or related to specific diseases are also often assigned an MIM number from McKusick's *Mendelian Inheritance in Man* (1986).

4. Methods Used for Gene Mapping

Probably the quickest and best way to map a gene is to "map by phone," by contacting a laboratory that is particularly devoted to gene mapping. A number of methods have been developed to map human genes. Some of these are described in detail below. Methods for DNA sequencing and "contig" mapping are not given in this brief review.

4.1. Genetic Linkage

Genetic linkage analysis is described in detail in following chapter, Genetic Linkage Analysis. This is the oldest technique for human gene mapping, and has been revitalized through the application of DNA studies. It is conceivable that, in the very near future, with the rapid expansion of genetic linkage analysis using

DNA probes in humans, standard large families can be used to map genes of interest very much as is now done with the recombinant inbred strains of mice. This will give information related to the recombination gene map that will then need to be correlated with physical mapping. A limited form of genetic linkage analysis is that of candidate gene linkage. In this situation, one utilizes biological/biochemical information to identify a gene that may be related to a specific disorder and to carry out linkage analysis in an affected family in search of cosegregation of the disease gene with the candidate gene. One may also use gene-map information to identify candidate genes based upon their position in the genome, and assignment of a diseased gene to a similar area on the chromosome by previous standard linkage analysis or through comparison of homologies between human and nonhuman genomes and hereditary diseases such as in the mouse.

Increasingly, gene-mapping studies are being done with DNA probes for specific genes. Physical mapping of these has been, and probably will continue to be, done mostly by the use of somatic-cell hybridization and *in situ* analysis, as described in detail below. For the physical mapping, many people feel that it is probably desirable to use both methods, which can be confirmatory in terms of chromosome assignment as well as giving regional assignment, particularly with the use of *in situ* studies.

4.2. Somatic-Cell Hybridization

Somatic-cell hybridization has been very successful in mapping human genes. With this approach, one utilizes interspecific somatic-cell hybrids formed as follows. To map human chromosomes, normal human cells, such lymphocytes or fibroblasts, are fused with an established rodent cell line. Usually, the rodent cell line has a deficiency in an enzyme, such as HPRT, which can be used to select against the rodent parent cell following hybridization. When human fibroblasts are used, the human parental cells can be selected against by the use of ouabain. The fused cells are then grown in culture. With growth, human chromosomes are lost from the somatic-cell hybrids in an approximately random fashion. This results in the formation of somatic-cell hybrid clones, which have different human chromosome content. Correlation of the presence of a given human gene with the human chromosomes that are left in the different hybrids permits one to map a gene to a specific human hybrid.

In addition to the loss of human chromosomes from the somatic-cell hybrids, the ease with which one can distinguish human from rodent genes has made this method particularly powerful, because it does not depend upon polymorphic variations among humans as has been necessary for the standard linkage analysis. Initially, many isoenzymes were mapped through the use of electrophoresis. However, at the present time, almost all of the mapping now utilizes DNA probes. This is not only because of the ready availability of an increasing number of DNA probes, but also because the DNA approach is independent of the need for gene expression to determine the presence or absence of a given gene. Indeed, DNA probes whose function is unknown can be mapped with this approach. Furthermore, nonhuman DNA or genes can be used for human mapping because of the high evolutionary conservation of many functional genes.

The somatic-cell hybridization technique can also be used to regionally map genes on chromosomes. This is generally accomplished through the use of human cells containing translocations that are fused with the rodent cells. The regional mapping is carried out by correlating the presence of a human gene with that part of the human chromosome that remains in the somatic-cell hybrid; this regional mapping is usually pursued after the gene has been assigned to a chromosome through the use of a general clonal somatic-cell hybrid panel. To increase the efficiency of obtaining hybrid clones that contain or lack the chromosome of interest, naturally occurring or introduced selectable markers may be utilized.

The following is a description of the method that we have found useful for the somatic-cell hybridization approach for mapping. As with most laboratory methods, each laboratory develops its own specific procedures or modifications of other procedures. Thymidine-kinase-deficient mouse cells can be used as the rodent parent; we have found mouse cells B82 (GM 0347A) to be useful. Normal human fibroblasts can be used; a well-established human fibroblast strain (preferentially male) is desirable, and we have found the IMR 91 strain to be useful. Both of these cultures can be obtained from the Mutant Cell Repository in Camden, NJ. The cells are fused in a mixed monolayer using a 50% solution of polyethylene glycol (mol wt 1000) in balanced salt solution (Davidson and Gerald, 1976). Approximately 24 h later, the monolayer of cells is divided into multiple independent culture dishes containing HAT medium (Szybalska and Szybalski, 1962; Littlefield, 1964) and ouabain (Kucherlapati et al., 1975). When the human cells are lymphocytes, the ouabain is not used. Multiple independent hybrid clonal cultures are isolated and a preliminary cytogenetic study can be done on ten metaphases to obtain an indication as to the human chromosome content of each clone. A panel of clones is then selected so that, using a combination of the clones in the panel, each and every human chromosome can be included or excluded for chromosome assignment when the panel is taken altogether. Once this clonal panel is established (often requiring 15–20 hybrid clones), they are grown to obtain enough cells for analysis, such as for isoenzymes or DNA. An aliquot is taken from each grown clone to analyze for chromosomes. Generally, a minimum of 30 banded metaphases is necessary per hybrid clone to establish the average chromosome makeup of that clone.

Each of the clones of the hybrid panel is analyzed for the gene of interest. Generally, if one is evaluating isoenzymes, fresh cells are required and a detection system specific for each enzyme is required. For DNA analysis, the DNA will need to be extracted. This can be accomplished as follows (modified from Fodor and Doty, 1977; Maniatis et al., 1982). Cells are suspended in 20 vol of 150 mM NaCl, 10 mM Tris, 10 mM EDTA, pH 7.9, 0.5% SDS, and 200 µg/mL proteinase K. After incubation at 37°C overnight, the samples are extracted twice with phenol-chloroform-isoamyl alcohol (25:24:1) and once with chloroform/isoamyl alcohol (24:1). The DNA is then precipitated with ethanol, washed, and redissolved in 10 mM Tris, 2 mM EDTA, pH 7.8 treated with RNase (40 µg/mL for 1h) followed by phenol extraction and ethanol precipitation.

The probe to be used for the hybridization has to be radioactively labeled. This is generally accomplished using ^{32}P to a specific activity of 1×10^8 cpm/µg by either nick translation (Rigby et al., 1977) or by random priming (Feinberg and Vogelstein, 1983).

The next step is to identify a restriction endonuclease enzyme that will distinguish the human from the nonhuman gene. We have found that, in most instances, either EcoRI or HindIII restriction endonucleases will permit distinction; if these two enzymes do not work, then it is a matter of having to try others until differences are found. These enzymes are used to cut the DNA from the parental cells that were used to set up the somatic-cell hybrids. This DNA then needs to be electrophoresed as follows. Approximately

10 μg of DNA that has been cut with the restriction endonuclease is electrophoresed through a 1.2% agarose gel in TAE (40 mM Tris acetate, 1 mM EDTA) buffer and then transferred by blotting to nitrocellulose or similar medium by the method of Southern (1975). This filter is then prehybridized for 12–16 h in 50% formamide, 5× SSC, 5× Denhardt's solution, 50 mM NaH$_2$PO$_4$ pH 6.5 and 250 μg/mL denatured salmon sperm DNA, at 42°C with shaking. The radioactive labeled probe is then hybridized to this filter in 45% formamide, 4.6× SSC, 5× Denhardt's solution, 20 mM NaH$_2$PO$_4$, pH 6.5, 250 μg/mL denatured salmon sperm DNA, and 10% dextran sulfate for 16 h at 42°C with shaking. The filter is then washed twice, first in 2× SSC, 0.1% SDS, and then in 0.1× SSC, 0.1% SDS, each time for 20 min at 55°C. The filter is dried briefly and exposed to Kodak XAR-5 X-ray film for a minimum of 2 d. The film is then developed, and the radioactive banding pattern is interpreted. For the parental cells, one needs the banding patterns of the two parental cell lines to be different. Once it is determined which restriction endonuclease gives this distinction, the DNA from the somatic-cell hybrid clonal panel is cut with the same enzyme and processed as described above. The autoradiograph is then interpreted for which hybrids contain the human gene and the correlation is made to determine on which chromosome the human gene is located.

4.3. In Situ Hybridization

In situ hybridization is playing an increasingly important role in mapping genes and regionally mapping them on human chromosomes. With this technique, a labeled DNA probe is hybridized directly to chromosomes, and a search is made for the location of the labeled probe on the human chromosomes. Most often, this technique is used in conjunction with the somatic-cell hybridization approach, and the two combined can give ready confirmation of the chromosome location of a gene.

Most often, the probe that is used is a cDNA probe; a genomic probe may contain repetitive sequences and may hybridize to many nonspecific chromosome sites. In general, it appears that, for the technique to be effective, the probe should be at least 500 bases in length. If there is hybridization to more than one site, this may indicate the presence of pseudogenes or homologies with some other gene, but this technique will not permit determination of which is the functional gene. The resolution of this technique has been estimated to be limited to about 10,000 kilobases or about 10 centimorgans.

Although there have been prior attempts to utilize *in situ* mapping, it was not until the report by Harper and Saunders (1981) that this became a reliable technique. As these authors noted that a major change in their approach was to utilize 10% dextran sulfate, which greatly accelerated the hybridization of DNA. For this technique, normal human chromosome metaphases are analyzed. The slides are generally used within a week of the time they are prepared. The slides are treated with RNase A in 2× SSC for 60 min at 37°C, washed free of RNase several times in 2× SSC, and dehydrated through alcohol series. The chromosomes on the slides are denatured in 70% formamide/2× SSC at 70°C for 2 min. The probe mixture is a pH 7.0 and contains 10% dextran sulfate sodium, 2× SSC, 50% deionized formamide, 200 µg/mL salmon sperm DNA, 2% Denhardt's solution. This is boiled for 5 min, frozen, and after thawing, is placed on slides under cover slips. The probe is labeled with tritiated nucleotides to a specific activity of approximately 3×10^8 CPM/µg DNA, using nick translation or oligolabeling as described above; approximately 5 ng of labeled probe are added per slide. The edges of the cover slips on the slides are sealed with rubber cement. Incubation is carried out in a tightly closed plastic dish for approximately 18 h at 37°C. The slides are then washed for 3 min in 50% formamide/2× SSC and 4 times for 2 min in 2× SSC, all at 39–40°C. For the detection of the radioactivity, the slides are dipped for 1 s in a liquid photographic emulsion (NT 92 Kodak/water at 42°C). After air drying in the dark for 2 h, the slides are placed in a light tight box with Drierite for 5 d at 4°C; longer or shorter times may be necessary for some probes. The slides are developed in Kodak Dektol fixed, washed, and allowed to dry for 24 h. The slides can be G-banded. The autoradiographs are directly analyzed from the microscope, although photographs can be made. Grain locations are recorded on a schematic representation of G-banded metaphase chromosomes, and all grains either on or touching a chromosome are scored. Depending upon the amount of autoradiographic labeling, probably a minimum of 50, and more likely more than 100, cells will need to be analyzed in order to obtain the overall distribution of the autographic grains on all of the chromosomes.

References

Davidson R. L. and Gerald P. S. (1976) Improved techniques for the induction of mammalian cell hybridization by polyethylene glycol. *Som. Cell Genet.* **2**, 165–176.

Feinberg A. P. and Vogelstein B. (1983) A technique for radiolabeling DNA restriction endonuclease fragments to high specific activity. *Anal. Biochem.* **132**, 6–13.

Fodor E. J. and Doty P. (1977) Highly specific transcription of globin sequences in isolated reticulocyte nuclei. *Biochem. Biophys. Res. Comm.* **77**, 1478–1485.

Harper M. E. and Saunders G. F. (1981) Localization of single copy DNA sequences on G-banded human chromosomes by in situ hybridization. *Chromosoma* **83**, 431–439.

Kucherlapati R. S., Baker R. M., and Ruddle R. H. (1975) Ouabain as a selective agent in the isolation of somatic cell hybrids. *Cytogenet. Cell Genet.* **14**, 362–363.

Littlefield J. W. (1964) Selection of hybrids from mating of fibroblasts in vitro and their presumed recombinants. *Science* **145**, 709–710.

Maniatis T., Fritsch E. F., and Sambrook J. (1982) *Molecular Cloning: A Laboratory Manual.* Cold Spring Harbor Laboratory, New York.

McKusick V. A. (1986) *Mendelian Inheritance in Man. Catalogs of Autosomal Dominant, Autosomal Recessive and X-Linked Phenotypes.* Seventh ed. The Johns Hopkins University Press, Baltimore.

Rigby P. W. J., Dieckmann M., Rhodes C., and Berg P. (1977) Labeling deoxyribonucleic acid by high specific activity in vitro by nick translation with DNA polymerase I. *J. Mol. Biol.* **113**, 237–251.

Southern E. M. (1975) Detection of specific sequences among DNA fragments separated by gel electrophoresis. *J. Mol. Biol.* **98**, 503–517.

Szbalska E. H. and Szbalski W. (1962) Genetics of human cell lines IV. DNA mediated heritable transformation of a biochemical trait. *Proc. Natl. Acad. Sci. USA.* **48**, 2026–2034.

Methods for Genetic Linkage Analyses

M. Anne Spence

1. Introduction

Reports of the mapping and cloning of genes appear regularly in the major newspapers and lay magazines. The catalog of mapped and cloned human genes is being produced biannually as the report of the Human Gene Mapping workshops. In the latest report (McAlpine et al., 1988), over 1200 genes were listed, of which half were cloned and over 2400 restriction-length polymorphic locations were identified in the human genome. The method of genetic linkage analysis goes hand in hand with the mapping of the genes and often is the key prior step in successfully mapping a locus of interest. Whereas mapping places a gene's location on a particular chromosome (often with sufficient precision to lead to application of cloning technology) linkage analysis defines the linear order among a series of loci. For a complete discussion of mapping techniques and methodology, see the chapter by Sparkes, this volume.

In this chapter, we will restrict ourselves to a description of the methods and problems associated with genetic linkage analyses. The presentation is divided into two sections; first is the discussion of the methodological concerns, which include both detection and estimation of linkage, and second is the discussion of the problems encountered after linkage is detected, which include genetic heterogeneity and gene order.

2. Basic Gene Linkage Analyses

During meiosis, homologous chromosomes pair for the first division, which will result in the reduction of the number of chromosomes to one-half. During the process, the synapses or pairing is quite precise at the DNA level. At this time in meiosis, physical breakage may occur in the strands of DNA and the strands from the two chromosomes become reattached to one another. Genetic loci, genes, are arranged in linear order along the chromosomes. For many of the loci, the two chromosomes carry different alleles, that is alternate forms of the gene. When the physical exchange occurs between the two chromosomes, these alleles are exchanged with respect to the loci on either side of their position. The exchange can be seen genetically as a new combination of alleles occurring in the next generation, and the general phenomenon of exchange is referred to as recombination. There are interesting theories about the recombination process, e.g., is it random, is it mandatory, what initiates the break, and so forth. Excellent texts exist that discuss the molecular aspects of the phenomenon (*see* for example *Genes III* by Lewin or the review by White and Lalouel, 1987). It is important here to realize that the molecular aspects of recombination are essentially irrelevant, since it is the *genetic* consequences of the exchange that we use in linkage analyses. Note that linkage was a major genetic tool in the hands of the Drosophila geneticists 25 years before we knew that DNA was the material of inheritance!

To see the genetic consequences of recombination, we must first define two genetic loci, which we will call locus A with alleles big A and little a, and locus B with alleles big B and little b. Since we are dealing with two members of a homologous pair of chromosomes, they by definition carry the same genetic loci and each chromosome would have a locus A and locus B. For example, on the maternal chromosome, the alleles present to mark the loci could be A and B, whereas on the paternal chromosome, the alleles present to mark the loci could be a and b. The gametes produced by this doubly heterozygous individual for the next generation would be those carrying A–B and those carrying a–b. If a recombination occurs between locus A and locus B, then two of the alleles exchange places. This recombination is referred to as crossing-over. The individual will continue to produce nonrecombinant (or parental) gametes of the type A–B and a–b from those meioses with no recombination, but will produce the *recombinant* gametes A–b and a–B from meioses where crossover has occurred. The data support the idea that recombination represents the break and exchange between a chromatid of each of the chromosomes. The

Linkage Analyses

frequency of recombination between two loci is a function of the distance between them. The closer two loci are located on the chromosome, the less likely it is that a recombination can occur between them, and the further apart they are located, the more likely it will occur. The measure of the *genetic* distance between two loci then is defined as the number of recombinant offspring out of the total number of offspring (the recombination frequency is usually designated by Greek θ). If the two loci are far enough apart on the same chromosome, recombination always occurs, and the recombinant and nonrecombinant gametes are produced in equal numbers. The recombination frequency for two loci sufficiently far apart is 50%. (Convince yourself that 50% recombination is the expected when the two loci A and B are on two different chromosomes and, thereby, must assort independently.) A recombination frequency of less than 50% indicates that two loci are not assorting independently and therefore are said to be linked.

Two obvious experimental restrictions appear immediately if we are to use linkage analysis. First, we must have two generations (at least be able to count gametes) in order to note recombinants, and second, we must have heterozygous individuals, where the alleles at each locus differ, permitting detection of crossover events. This is the clearest separation between linkage, which has these requirements, and genetic mapping (*see* chapter by Sparkes), which does not.

The question of determining recombination frequency is relatively simple for experimental organisms like *Drosophila* or mouse where these two restrictions are readily met. Matings can be constructed so that the genotypes of the parents are known exactly and offspring from many such matings can be directly counted for parental and recombinant phenotype.

The importance of the experimental restrictions becomes fairly obvious when one observes the usual material available for analysis in human data. The first restriction of at least two generations can often be met, except when the doubly heterozygous parent is no longer available for study (for example, when the parent carrying the gene for a dominant progressive disorder like Huntington's Disease has died). The second restriction of knowing phase to count recombinants is a more serious one. For locus A and locus B, the phase could be in *coupling* with alleles A and B on the same chromosome (and likewise a and b), or in *repulsion* with A and b on one chromosome and a and B on the other. Phase is often given in short hand by AB/ab or Ab/aB. The obvious importance of phase is that the recombinant offspring of one phase would be the parental (nonrecombinant) type for the other phase. The terms coupling and repulsion are convenient when dominant and recessive alleles are present, but ambiguous when any locus is codominant. However, the impact on the linkage analysis is the same. Many of the families available for linkage studies in humans are only two or partially three generations, and the phase of an individual cannot be determined. The rest of this section will deal with handling the methodological problems in human data, since in very large families with phase clearly specified, one could in theory count recombinants as one could with experimental data and as the first linkages were identified in humans (Mohr, 1954).

To deal with the methodological problems of sample size and independent assortment, a statistical approach for analyzing linkage data in humans was proposed in 1955 by Morton and is the standard approach today. Some limitations and possible changes are mentioned briefly in the summary below. The idea was to analyze the information present in each family, and then to add them together in a meaningful way to have sufficient information to detect linkage and to estimate the recombination frequency. To accomplish this, a lod score (or *log* of the *od*ds) is calculated for each family separately. The definition of a lod score is relatively simple, and the calculation becomes complex only as family sizes become large or burdensome as the number of loci becomes large. Most lod scores are calculated by using one of the readily available computer algorithms, but some understanding of the concept is important for interpreting the computer results.

To calculate a lod score, one must calculate the probability of observing a family given information about the individual's genotypes (or most often phenotypes). If you were asked the probability of a family with two children whose phenotype were male, the answer would be a .5 × .5 or .25. It is as simple to compute the probability of observing the family given below:

(father) A–B/a–b   (mother) a–b/a–b
(child 1) A–B/a–b   (child 2) A–b/a–b

and we must do it two times: (1) for the two loci *if* they are linked with (θ) equal to 10%, which would be nonrecombinant times recombinant or .90 times .10, and (2) for the two loci if they are

Linkage Analyses

assorting independently (θ = 50%), which would be just .5 × .5. The lod score of a family is defined as the log to base 10 of the ratio

$$\frac{\text{Probability (family } \theta = 10\%)}{\text{Probability (family } \theta = 50\%)}$$

The log is taken so that the lod scores may be added. Since each family is an independent observation, their probabilities would be multiplied, but the numbers become small and difficult to handle. The calculation could be done for any value of recombination, since *a priori* we do not know if locus A and locus B are linked. An attempt has been made to standardize the scores for purposes of reporting and pooling over studies, and the usual values calculated are for θ = .001, .01, .05, .1, .2, .3, .4, and of course, .5. The lod score could also be calculated for this family if the phase of the father were not known, which is the most likely case given the two generation information above. The numerator is calculated twice, once with the father in coupling and once with him in repulsion, and the average is taken (sum the two scores and divide by two), since the father must be either in coupling or repulsion and they are equally likely *a priori*. The phase does not affect the calculation of the denominator, since there the two loci are assumed not to be linked. This step reduces the power of the analysis, but deals with the problem of unknown phase in a statistically unbiased manner.

A single small family such as this one would have a small lod score, even though phase is assumed for the father, since there are only two children. It is by adding scores together over small families that *statistically significant* evidence for linkage can be obtained and the problem of small sample size for human families overcome. If the two loci are truly linked at about θ = 10%, then the numerators will *tend* to be larger than the denominators, since that is the "true" probability and the lod scores will tend to be positive and the sum will grow larger. By convention, a lod score of 3.0 or greater is taken as statistical evidence of linkage (one where the numerator is 1000 times more likely than the denominator). Also by the same general rule, if the loci are not linked, the denominators will tend to be larger and the log of a number less than one is negative. Therefore, when a lod score hits –2, it is considered evidence against the linkage of the two loci *at the value of θ for which the scores were calculated*. It is extremely important to emphasize that the scores will tend to be in the true direction, but it is not the case that they will always be positive for each family even when the two loci are closely linked. For example, if there were only one recombinant child in every 100 offspring, that recombinant could occur in a two-child family where the family would appear to assort independently and, therefore, give a negative lod score. We will return to this point when we discuss the problems encountered after linkage is detected.

Once the principles of linkage and lod scores are understood, then it is time-efficient and more accurate to have a computer algorithm do the repetitive calculations. Several standard programs are readily available, including LIPED (Ott, 1974) or LINKAGE (Lathrop et al., 1985). There are, however, four areas where the calculation of lod scores is subject to substantial error, if the parameters are not correctly specified when the data are entered. We will briefly define and discuss each of these potential problem areas. There are other concerns about the more detailed interpretation of lod scores and linkage analyses. Excellent references for more detail are Ott (1985) or White and Lalouel (1987).

The first problem area is the assignment of the mode of inheritance for the loci being analyzed. This is quite straightforward for blood groups, serum enzymes, or most RFLPs. However, it may present a major stumbling block if one of the traits of interest is a human disease for which the mode of inheritance is unknown or controversial. In order to do the probability calculation, a mode of inheritance must be specified, which means stating the relationship between the observed phenotypes of the individuals and the alleles segregating at that locus. Changing the mode of inheritance can make a substantial difference in the value of the calculated lod score (Cox et al., 1988; Hodge 1983). It is easy to visualize this problem if one simply thinks of the following family:

(father) A–B/a–b   (mother)a–b/a–b

If the disease is caused by the dominant allele A, then the father is affected and the family contributes to the lod scores. However, if the disease is caused by the recessive allele (a), then the mother would be affected and the lod score would be zero since she is not doubly heterozygous. Theoretical work is being conducted on how to handle this problem of diseases that do not have clearly defined Mendelian inheritance patterns.

The second area of concern arises when there are individuals in the pedigree who have not been studied for whatever reason. These individuals must be coded "unknown." The algorithms are set up to do the calculations over all possible genotypes, weighting the final lod score by Hardy-Weinberg frequencies. Part of the setup of the analysis is to supply gene (allele) frequencies. If there are any unknown individuals in the pedigree, the gene frequencies used may substantially affect the absolute value of the lod scores.

For an example from our laboratory, the assignment of the correct equal frequencies to two alleles of an RFLP segregating in a three-generation pedigree with X-linked recessive Norrie's disease resulted in a maximum lod score of 1.76 at $\theta = 10\%$. If one allele was designated rare, i.e., frequency = .001, the maximum lod score still occurred at 10% recombination, but the value changed to 2.06. Reversing the rare assignment to the second allele resulted in statistically significant evidence *against* linkage at 5% recombination and no positive lod scores at all. This dramatic change resulted because a single male (father of the informative female) had not been typed. Although extreme, the example clearly illustrates the potential sensitivity of the scores to gene frequencies.

The effect of different frequencies also holds when the "markers" are coded as haplotypes and is often a more serious problem, since individual frequencies may be known for the loci but not for the haplotypes. Closely linked loci analyzed as haplotypes to increase the number of informative people are often so close together that they are not in Hardy-Weinberg equilibrium, and the use of assumed equilibrium frequencies may not be correct.

The third area for potential problems is where the phenotype–genotype relationship is not strictly one-to-one, the most common reason being a delayed age-of-onset in individuals who are at risk to develop a disease. To call an individual age 10 unaffected for Huntington's disease even though the mother is affected could seriously bias the lod scores (Hodge et al., 1979). In situations such as this, an age of onset correction must be applied in the analysis. Such corrections are usually available as part of the packaged computer program. The exact algebraic form of the corrections is not important (Hodge et al., 1979), but the investigator must have some information about the onset of the disease in the relevant population in order to set some boundaries for the calculations.

The fourth concern is simply the fact that the recombination frequency does not appear to be the same in the two sexes and may in fact differ strikingly even within the same region of the genome (Rao et al., 1979). The first estimates of θ led us to believe that male recombination occurred less often than female recombination (also noted in experimental organisms). The human map distances were defined in terms of male recombination frequencies. However, more recent data suggest that, in some regions of the genome, the female rate may be lower than the male rate and the frequencies may vary from chromosome to chromosome and chromosome region to chromosome region (Meyers and Beaty, 1986). The important consideration for the calculation of lod scores is that the maximum lod score often does not occur at a value of $θ_{male} = θ_{female}$. The existing programs provide the option to calculate a matrix of recombination frequencies with the frequency equal as well as not equal between the sexes. Such matrices facilitate the pooling of data across studies where the maximum has been found to be off the diagonal (i.e., $θ_{male}$ not equal $θ_{female}$).

The preceding discussion focuses on the problems of detecting linkage, that is of calculating lod scores in family to achieve a total lod score of greater than 3.0. The direct consequence of detecting linkage is the estimate of the recombination frequency or the genetic distance between the two loci. When lod scores are used for the analysis, then the maximum lod score occurs at the maximum likelihood estimate of the recombination frequency. Computer algorithms are available to provide interpolated estimates of θ, even when the recombination frequency has been treated separately for the two sexes (for example, see QUAD in Hodge et al., 1983).

The accuracy of the estimate of recombination is important in the assembling of genetic maps of numerous linked loci and is critical in the application of linkage data to prenatal counseling. One important hypothesis that may be tested in the linkage studies is the "candidate gene" hypothesis, where the null hypothesis is θ = zero, i.e., the locus is the gene for the disease of interest. This approach has assumed an important role with the cloning of many of the structural protein genes. Now we can ask a question of the type: is it a defect in the γ crystalline gene that results in a specific form of childhood cataracts (Barrett et al., 1988)? If the cloned gene is responsible for the disease, there should be virtually no recombination between the clone detected RFLP and the occurrence of the disease. This approach is being applied on a grand scale for the skeletal disorders, such as osteogenesis imperfecta types I to IV and the collagen genes (Spence and Tsui, 1988) and is relevant for the neurological disorders as well. Relatively small samples can be very powerful under this approach, since a single confirmed crossover negates the hypothesis.

3. After Linkage is Detected

The detection of linkage is only the first step in using linkage relationships to further our understanding of the genes involved. The direct consequences can be as straightforward as confirming the existence of a gene in traits where the genetic etiology was uncertain, such as manic-depressive psychosis (Egeland et al., 1987) or juvenile onset, insulin-dependent diabetes (Hodge et al., 1983), or can be as puzzling as the cloning and dissection of the huge and complex Duchenne muscular dystrophy gene (region?) on the X chromosome (Koenig et al., 1987). We will discuss two problems that must be addressed after the lod scores are statistically significant; first is the issue of genetic heterogeneity, and second is the problem of multiple loci or multipoint mapping.

Genetic heterogeneity is the situation encountered when a single phenotype results from the segregation of alleles at more than one locus. In the first direct application of lod scores in 1956, Morton demonstrated that 80% of the families segregating elliptocytosis were linked to the RH blood group locus and the remaining families segregated independently of RH. Subsequent to that finding, the biochemists confirmed that the disease that appeared homogeneous by clinical criteria is caused by different biochemical mechanisms and independent genetic loci (Conboy et al., 1986). The discovery of genetic heterogeneity is an important consequence of linkage analyses and is more readily detected as the map becomes more complete. For example, we now know that the clinical separation of osteogenesis imperfecta into subtypes does not correspond with the linkage to the various collagen loci (Spence and Tsui, 1988). However, when the Huntington's disease gene was mapped, there was interest in whether or not the different appearing forms of the disease, i.e., early onset, rigid, and so on, were the result of different loci. To date, all families studied are consistent with the linkage relationship of a single locus on the short arm of chromosome 4 (Haines et al., 1986). The problem is to determine when genetic heterogeneity (vs clinical variability) exists. To this end, there must be a statistical test to determine whether the lod scores vary simply because of chance segregation (as discussed above) or whether they differ because they actually represent two separate loci, one linked and the other unlinked. Several types of heterogeneity analyses have been proposed that test for statistically significant differences among the scores.

Morton (1956) proposed a predivided sample test where the families could be divided into two subgroups based on biochemical or clinical criteria. The statistic is calculated over the pooled scores and approximates a χ square distribution. Another approach is to estimate simultaneously the proportion of linked families denoted α and the proportion unlinked $(1 - \alpha)$, testing if α is different from zero (Hodge et al., 1983). Neither approach has much power if the families are small, resulting in scores of small absolute values.

As the linkage map grows, groups of linked loci are analyzed as a unit in an approach known as multipoint mapping. The first objective of these analyses is to establish the order of the loci on the chromosome. For a series of loci a reasonable distance apart, say 20% recombination, the order can be easily established by comparison of the pairwise distance estimates and the observation of double crossover events. For a series of loci located within a small region, the pairwise estimates of distance will overlap considerably, since they are relatively imprecise and may not distinguish any one order. Within such small regions, double crossovers are very rare events and may not be observed to assist with the ordering. In this situation, the loci are analyzed as a set to estimate simultaneously order and map distances, a process referred to as multipoint mapping. The method does have more power to resolve the question of order than just two-point mapping alone (Bishop, 1985), but is not guaranteed to produce an unequivocal result. Using data from two laboratories, numerous groups of investigators attempted to determine the order of loci on the short arm of chromosome 11 for an exercise at a Genetic Analysis Workshop (Meyers and Beaty, 1986). Although there was consensus on the general order, several loci, namely insulin (INS) and HRAS, could not be ordered unambiguously. The same problem arose at the Human Gene Mapping 9 Workshop when investigators presented their data on the region surrounding the cystic fibrosis locus on chromosome 7 (Spence and Tsui, 1988). Computer algorithms

Linkage Analyses 173

are available that do multipoint mapping (Lathrop et al, 1985; Lange, personal communication). Here the mathematical analyses become more complex as the possible orders are compared on the basis of the likelihood of their fit to the data. The accepted standard is to report the best fitting order and at least the first alternative including the difference in likelihood achieved by the best fit.

4. Summary

In summary, linkage analysis is a powerful genetic tool that opens the door to many possibilities of further exploration of the genes of interest. Complex statistical calculations are required to detect linkage in human data because

1. phase is often unknown or key individuals are not available for study, greatly reducing the amount of information
2. most families are too small to be definitive in and of themselves
3. for more loosely linked loci, it requires a great deal of data to distinguish linkage from the null hypothesis of independent assortment
4. genetic heterogeneity may result in families that are linked to a marker mixed with families where the locus of interest is not linked and
5. age of onset or other forms of reduced penetrance must be accounted for in the analyses.

The definition of a lod score is straightforward, but the calculation with its related assumptions is nontrivial. It is relatively simple to submit data to the existing programs in a form that produces lod scores, but the correctness of those scores depends to a large degree on a number of input factors such as allele frequency, specified mode of inheritance, and the penetrance function.

The future is to use multipoint mapping as the number of linked and mapped loci continues to grow exponentially (McAlpine et al., 1988). The multipoint approach still begins with the calculation of two-point lod scores, but expands the potential for determining order for the loci and increases the power for detecting linkage for a new locus within an existing group of linked loci. The problem for the future that has not yet been resolved is how to present and preserve the phenomenal number of lod scores being generated. Several suggestions have been made for a condensed form of presentations (see, for example, MacLean et al., 1985), but no consensus has been achieved.

References

Barrett D. J., Sparkes R. S., Gorin M. B., Bhat S. P., Spence M. A., Marazita M. L., and Bateman J. B. (1988) Genetic linkage analysis of autsomal dominant congenital cataracts with lens-specific DNA probes and polymorphic markers. *J. Ophthal.* **95,** 538–544.

Bishop D. T. (1985) The information content of phase-known matings for ordering genetic loci. *Genet. Epidemiol.* **2,** 349–361.

Conboy J., Mohandas N., Tchernia G., and Kan Y. W. (1986) Molecular basis of hereditary elliptocytosis due to protein 4.1 deficiency. *New Engl. J. Med.* **315,** 680–685.

Cox N. J., Hodge S. E., Marazita M. L., Spence M. A., and Kidd K. K. (1988) Some effects of selection strategies on linkage analysis. *Genet. Epid.* **5,** 289–297.

Egeland J. A., Gerhard D. S., Pauls D. L., Sussex J. N., Kidd K. K., Allen C. R., Hostetter A. M., and Housman D. E. (1987) Bipolar affective disorders linked to DNA markers on chromosome 11. *Nature (Lond.),* **325** 783–786.

Haines J., Tanzi R., Wexler N., Harper P., Folstein S., Cassiman J., Meyers R., Young A., Hayden M., Falek A., Tolosa E., Crespi S., Campanella G., Holmgren G., Anvret M., Kanazawa I., Gusella J. and Conneally M. (1986) No evidence of linkage heterogeneity between Huntington's disease (HD) and G8 (D4S10). *Am. J. Hum. Genet.* **39, Suppl.,** A156.

Hodge S. E. (1983) Epistatic two-locus models: Bias in estimate of θ. *Am. J. Hum. Genet.* **35,** 1053–1054.

Hodge S. E., Anderson C. E., Neiswanger K., Sparkes R. S., and Rimoin D. L. (1983) The search for heterogeneity in insulin-dependent diabetes mellitus (IDDM): Linkage studies, two-locus models, and genetic heterogeneity. *Am. J. Hum. Genet.* **35,** 1139–1155.

Hodge S. E., Morton L. A., Tideman S., Kidd K. K., and Spence M. A. (1979) Age-of-onset correction available for linkage analysis (LIPED). *Am. J. Hum. Genet.* **31,** 761–762.

Koenig M., Hoffman E. P., Bertelson C. J., Monaco A. P., Feener C., and Kunkel L. M. (1987) Complete cloning of the Duchenne muscular dystrophy (DMD) c DNA and preliminary genomic organization of the DMD gene in normal and affected individuals. *Cell* **50**, 509–517.

Lathrop G. M., Lalouel J. M., Julier C., and Ott J. (1985) Multilocus linkage analysis in humans: Detection of linkage and estimation of recominbation. *Am. J. Hum. Genet.* **37**, 482–498.

Lewin B. (1987) *Genes III* (John Wiley New York).

MacLean C. J., Morton N. E., and Lew R. (1985) Efficiency of LOD scores for representing multiple locus linkage data. *Genet. Epidemiol.* **2**, 145–154.

McAlpine P. J., Van Cong N., Boucheix C., Pakstis A. J., Doute R. C., and Shows T. B. (1988) The 1987 catalog of mapped genes and report of the nomenclature committee. *Cytogenet. Cell Genet.* **46**, 29–101.

Meyers D. A. and Beaty T. H. (1986) Genetic analysis workshop IV: Summary of two-point and multipoint mapping of 11p. *Gene. Epidemiol.* **suppl. 1**, 99–111.

Mohr J. (1954) *A Study of Linkage in Man* (Munksgaard, Copenhagen,).

Morton N. E. (1955) Sequential tests for the detection of linkage. *Am. J. Hum. Genet.* **7**, 277–318.

Morton N. E. (1956) The detection and estimation of linkage between the genes for elliptocytosis and the Rh blood type. *Am. J. Hum. Genet.* **8**, 80–96.

Ott J. (1974) Estimation of the recombination fraction in human pedigrees: Efficient computation of the likelihood for human linkage studies. *Am. J. Hum. Genet.* **26**, 588–597.

Ott J. (1985) *Analysis Human Genetic Linkage* (The Johns Hopkins University Press, Baltimore).

Rao D. C., Keats B. J., Lalouel J. M., Morton N. E., and Yee S. (1979) A maximum likelihood map of chromosome 1. *Am. J. Hum. Genet.* **31**, 680–696.

Spence M. A. and Tsui L. C. (1988) Report of the committee on the genetic constitution of chromosomes 7, 8 and 9. Human Gene Mapping Workshop 9, *Cytogenet. Cell Genet.* **46**, 170–187.

White R. and Lalouel J.-M. (1987) Investigation of genetic linkage in human families. *Adv. Hum. Genet.* **16**, 121–228.

Lineage Analysis and Immortalization of Neural Cells via Retrovirus Vectors

Constance Cepko

1. General Introduction

Retrovirus vectors are extremely versatile tools that have been used in a wide range of systems. They have recently been employed in studies of developmental neurobiology and show great promise for future work in the neurosciences [see reviews by Price et al. (1987) and Cepko (1988a,b, 1989)]. Although several features contribute to the popularity and versatility of retroviruses, their most important attribute as vectors is their efficient integration system. This feature, which is generally unlimited in host range, has enabled the transfer of genes isolated from a wide range of host species into many different types of cells. Gene transfer can be accomplished in vitro or in vivo. The in vivo capacity offers a major advantage over other gene transfer techniques, such as transfection or electroporation, which are generally limited to in vitro use. A discussion of the basic virology that underlies vector functions, as well as potential applications and protocols follows.

2. Introduction to the Virus

Retroviruses are naturally occurring, RNA-containing, viruses. The single-stranded RNA genome is encapsidated by an icosahedral protein shell that is surrounded by a lipid bilayer (for a review of retroviruses, see Weiss et al., 1984, and 1985). Retroviruses are produced by a budding process carried out by the host cell. This is typically a nonlytic phenomenon and is innocuous with respect to the host. They have been isolated from every species in which a serious isolation attempt has been made. There are many examples of these viruses from mammals and birds. Fish retroviruses have also been reported, although these descriptions have been brief, and no attempts to create fish retrovirus vectors have been reported. To date, there are no recorded attempts to isolate retroviruses from amphibians. Since relatives of these viruses have also been observed in yeast and Drosophila, it is likely that such viruses exist in all species and that they can be adapted as vectors. An adaptation of the existing mammalian or bird retroviruses, as discussed below, may provide a wide host-range virus for many different organisms.

All retroviruses share the same basic life style. They enter the host cell through an interaction at the host cell surface. Presumably, there is a specific protein receptor on the host cell surface. There is a preliminary report of isolation of the gene for the ecotropic murine receptor (Cunningham et al., 1986). Although an extensive mapping of cells expressing this receptor has not yet been accomplished, it appears to be present on most—perhaps even all—cell types in a given host species. To gain entry to the host cytoplasm, there is an interaction between the host receptor and the viral envelope protein [a glycoprotein of 70,000 kdaltons for Moloney murine leukemia virus (MMLV)]. After entry, synthesis of viral DNA occurs within the cytoplasm. The viral genome encapsidated in the viral particle is in the form of two identical RNA molecules. These are reverse transcibed into DNA by the viral enzyme, reverse transcriptase, which is also encapsidated in the virion particle. The viral DNA is subsequently found in the nucleus in a double-stranded linear or circular form. The next event is the crucial element of the life cycle as far as the use of these viruses as vectors is concerned. The viral DNA reproducibly integrates into the host genome. It appears that there is a very large number of sites in the host cell genome for the majority of viral integration events, although a recent report by Shih et al. 1988 indicates that there are a number of host sites that are overrepresented as integration sites. The virus is quite precise concerning the site on the viral genome that is used for integration. It always integrates such that its own genome is uninterrupted, at a site at the tips of redundant viral sequences, called long terminal repeats, or LTRs. Thus, the integrated viral genome ("provirus") preserves all of the information it needs for expression of its genes. Integration into nonreplicating λ DNA in vitro has recently been demonstrated and should lead to insights into the mechanism of this process (Brown et al., 1987). This unique aspect of the retrovirus life cycle, that is, the efficient, precise, and stable integration event, is the key feature for their use in gene transfer. The integration event places at least one demand on the host cell. There must be an S phase in order for viral integration to proceed. The reason for this is unknown, but perhaps there is a requirement for activities of host enzymes that are only made during S phase. Integration into postmitotic neurons has never been observed, and may represent a serious limitation for certain types of experiments. However, it is possible to integrate the genome into the mitotic progenitors of neurons. The integrated viral genome passes to all daughter cells in a reliable, Mendelian fashion, usually with no rearrangements, deletions, or movement of viral sequences. A single copy of the provirus results from each competent, infectious particle.

Subsequent to integration, a wild type replication-competent provirus proceeds to make more retroviruses by expressing the viral structural genes. This is accomplished by a single strong promoter, located in the 5' LTR. Two (and possibly three) mRNAs result, and these serve to express the viral structural information. Assembly of the virus then occurs, and a complete, infectious particle buds from the host plasma membrane. Encapsidation of the viral genomic RNA occurs via a sequence, named ψ, located in the 5' end of the virus (and possibly extending into the gag gene), which is apparently recognized by the packaging proteins (Watanabe and Temin, 1982; Mann et al., 1983; Bender et al., 1987). RNAs that are lacking this signal (cellular mRNAs and env mRNA) do not become efficiently packaged. This entire process is quite benign. No perturbation of host cell functions has been observed. In fact, in order to find evidence of virus infection, one must specifically look for viral RNA, DNA, proteins, or budding particles.

3. Adaptation of Viruses as Vectors

3.1. General Vector Design Strategy

Adaptation of retroviruses as vectors proceeded in several laboratories with the advent of cloning techniques that permitted the manipulation of the viral genome as bacterial DNA plasmids. Proviruses were cloned from infected cells and manipulated with standard techniques to yield a variety of vectors. Unfortunately, the best vector for any particular gene or application is not always predictable. Vector designs are continually evolving, and the choices are sometimes bewildering. The best strategy is to obtain vectors from a laboratory that has experience with several different designs and that will advise on their use. The first vector design that is chosen may work beautifully. Alternatively, many vector designs may be tried and still only result with low transmissability and poor expression. In my experience, it is best to keep things as simple as possible. For general descriptions of vectors and transmission strategies, see Mulligan 1983, Chapter 4S in Weiss et al., 1985, and Brown and Scott, 1987.

Vectors contain the cis-acting viral sequences necessary for the viral life cycle. These include the ψ packaging sequence, reverse transcription signals, integration signals, viral promoter, enhancer, and polyadenylation sequences. A cDNA can thus be expressed in the vector using the transcription regulatory sequences provided by the virus. Alternate constructs that lack the viral promoter and/or enhancer have also been made (Yu et al., 1986; Hawley et al., 1987). These constructs are used for certain applications where one wishes to study regulation of an alternative promoter or avoid activities that may depress gene expression because of repression of the viral LTR. As vectors do not include the structural genes for production of a retrovirus particle, these functions must be supplied in trans, as discussed in the next section.

3.2. Packaging Lines and Methods for Virus Production

3.2.1. Rationale

How does one make an infectious viral particle from a bacterial plasmid that contains the retrovirus cis-acting sequences? The ability to make a virus particle containing a vector genome is provided by specific cell lines, called "packaging" lines. These lines are usually derived from mouse fibroblasts and contain all of the structural protein information that is necessary for production of retrovirus particles. They contain the viral gag, pol, and env genes as a result of the introduction of these genes by prior transfection. However, these lines do not contain the packaging sequence, ψ, on the viral RNA that encodes the structural proteins. Thus, the packaging lines make viral particles that do not contain the genes gag, pol, or env. Prior to the introduction of vector DNA into these lines, cellular RNAs are randomly encapsidated and budded as "normal" viral particles. When one wants to produce viral particles that contain the vector genome, the vector DNA is introduced via transfection or infection. Protocols for this purpose are given in section 3.2.3.

3.2.2. Host Range

The host range of a given virus appears to be determined by several factors. However, the present discussion will be restricted to the issue of entry into the host cell. The interaction between the glycoprotein on the virus envelope, a product of the viral env gene, and the host cell receptor is necessary for the productive entry of the viral particle into the host cytoplasm. The murine viruses have several classes of viral glycoproteins that interact with different host cell receptors. The most commonly used class of viral glycoprotein is the ecotropic class. These viral glycoproteins allow entry only into rat and mouse cells via the ecotropic receptor on these species. They do not allow infection of humans and, thus, are considered relatively safe for gene transfer experiments. The most frequently used packaging line that encodes the ecotropic env gene is called ψ2 (Mann et al., 1983). This was the first ecotropic packaging line, and it makes the highest titers of vectors, relative to other packaging lines (for unknown reasons). Recently, two new ecotropic packaging lines, CRE (Danos and Mulligan, 1988) and GP+E–86 (Markowitz et al., 1988) have been made. These new lines are designed to avoid the generation of helper virus (described below). The env gene of the amphotrophic class endows virus with a very broad host range, including mouse, human, chicken, dog, cat, and mink. There are four packaging lines for the production of vectors with this coat: ψam (Cone and Mulligan, 1984), PA12 (Miller et al., 1985), PA317 (Miller and Buttimore, 1986), and CRIP (Danos and Mulligan, 1988). The design for CRIP is similar to that of CRE and is, thus, the least likely of all of the amphotropic lines to lead to production of helper virus. There is also a packaging line, D17–C3, based on the avian virus, spleen necrosis virus (Watanabe and Temin, 1983). This line also produces a virus with a broad host range, which includes birds and mammals, and will encapsidate MMLV vectors (Embretson and Temin, 1987). In fact, all of the aforementioned lines will recognize the ψ sequence present on the MMLV vectors. They will thus encapsidate vectors that do not originate in the homologous class of virus that was used to make the packaging line. For example, the BAG genome, which was derived from MMLV (Price et al., 1987), was introduced into the ψ2, ψam, PA317, and D17–C3 lines by crossinfection. Reasonable viral titers were made by all packaging lines, with the ψ2 line producing the highest (10^6 CFU/mL) and the D17–C3 line producing the lowest (10^4 CFU/mL).

A stable packaging line, Q2bn, based on avian leukemia virus (ALV), was recently made by Stoker and Bissell (1988). This line was designed for production of ALV-based vectors and will not encapsidate MMLV-based vectors. The ALV-based vectors are currently being developed (for examples, see Hughes et al., 1987; Stoker and Bissell, 1988) and have lagged somewhat behind the MMLV-based vectors. It is not yet clear whether Q2bn and current ALV-based vectors will passage at as high a titer as the MMLV counterparts.

Introduction of the viral genome into packaging lines can be accomplished by transfection or by using a virus produced by one line to infect a different packaging line, as described below. Infection of a packaging line by a virus produced by the same packaging line is very low in efficiency (although it is not blocked entirely). The viral glycoprotein produced by a packaging line apparently binds the host cell receptor for that env glycoprotein class. A virion bearing that same env class thus cannot easily enter. However, since the different classes of viral glycoproteins use different cellular receptors, one can "cross-infect" by using virions produced by one class of packaging line to infect another (e.g., CRE-produced virus can infect CRIP cells quite effectively). There is also an alternative method that allows infection of a packaging line with virions produced by that packaging line class (Rein et al., 1982). One can use an inhibitor of glycosylation, such as tunicamycin, to block the production of the env glycoprotein transiently. During the application of this drug block, the cells can be infected with virions bearing the same class of env glycoprotein. The drug is then washed out. This method requires careful titration of the drug parameters (time, dose) as these drugs can be toxic to the cells.

Species that are classically used for embryological studies such as newts, frogs, and fish cannot be infected by the vectors available today. As mentioned above, there has been no isolation of retroviruses from amphibians, and the fish retroviruses have not yet been made into vectors. However, there is an ongoing effort to generate a packaging line, or a series of lines, that will allow infection of a very broad host range, including fish and amphibians. Cliff Tabin (Harvard Medical School, Department of Genetics and Massachusetts General Hospital, Department of Molecular Biology) and Richard Mulligan's laboratory (Whitehead Institute, MIT) are attempting to use other viral glycoproteins in place of the retroviral env products to allow the retrovirus to enter virtually any cell. It appears from preliminary experiments (Tabin, personal communication) that the viral functions subsequent to entry (uncoating, reverse transcription, integration, transcription from the viral LTR, and translation and activity of β-gal) can occur in newt cells.

3.2.3. Methods of Virus Production

Viral genomes can be introduced into packaging lines via transfection or infection. Transfection is a nonspecific entry technique that allows one to introduce any DNA into some types of cells (see protocol 3.2.3.1.). This DNA can be transcribed transiently from about 10% of the cells, within the period of a few hours to a few days after transfection. However, only about 1 in 5000 cells will stably integrate the DNA. The transiently or stably produced vector RNA (which contains ψ) is efficiently packaged and budded from the surface of the packaging cell line. Thus, the infectious retrovirus vector can simply be harvested by removal of the supernatant of the packaging line. A given line that has stably integrated the virus may produce high amounts of virus indefinitely. Cells that have stably integrated the viral genome are usually selected by application of a drug that selects for a gene encoded by the virus. If no such dominant, selectable gene is encoded by the virus, the transfected cells can be cotransfected with a nonviral plasmid that does encode such a gene. By using a molar ratio of viral:nonviral plasmid of 10:1, one can select for cells that have integrated both plasmids. If one is generating a producer by cross-infection, a very small amount of virus should be used for the infection. If a high multiplicity of infection (moi) were used one runs the risk of transferring defective genomes or wildtype helper recombinants (discussed in the next section).

The amount (titer, expressed as colony forming U/mL, CFU/mL) of virus produced varies widely, and can be as high as 10^8 CFU/mL (see protocol 3.2.3.2.). Typically, 10^6 CFU/mL is considered to be a good titer for an ecotropic producer. Viral stocks can also be concentrated by centrifugation (see protocol 3.2.3.) or polyethylene glycol precipitation (Aboud et al., 1982). Viruses produced by packaging lines can be used to transduce the viral genome by simply incubating the virus with the cells to be infected.

For most applications in vitro, a polycation, such as polybrene or DEAE-dextran, is used to aid in viral infection. These polycations apparently can promote virus binding to the host cell surface by reducing electrostatic repulsion between the negatively charged surfaces of the cell and virion. Alternatively, the cells to be infected can be incubated with the packaging line (cocultivation method; *see* protocol 3.2.3.4.) This method is used to infect hematopoetic cells and appears to increase the infection efficiency greatly (Williams et al., 1984; Lemischka et al., 1986). It is difficult to generalize about the efficiency of infection, although it can be close to 100%. The variables that influence the efficiency of infection of different cells are unknown.

3.2.3.1. TRANSFECTION FOR VIRUS PRODUCTION. This protocol is from Parker and Stark (1979) and is a modification of the $Ca_3(PO_4)_2$ transfection protocol of Graham and Van der Eb (1973). Filter sterilize the solutions, and work in a sterile hood.

1. Plate packaging cells to about 10–20% confluency (hereafter referred to as a 1 → 10 or 1 → 5 split) the day before. This protocol is based on the use of a 10-cm dish, but one can use a 6-cm dish and scale everything down by a factor of 3.
2. Place 10 μg of the retrovirus vector plasmid DNA into 0.5 mL HBS (*see* formula below) and mix. The DNA need not be sterile. We use Falcon tubes, #2054, for good visibility of the precipitate. If you are doing a cotransfection, use a molar ratio of viral to nonviral (or drug marker) plasmid of 10:1. Mix them together in the HBS.
3. Add 32 μL of $2M$ $CaCl_2$ while gently shaking the tube. Tap the tube for about 30 s.
4. Incubate at room temperature for 45 min. A fine, hazy blue precipitate should develop. Large, clumpy precipitates do not work well.
5. Remove medium from cells and gently pipet the HBS-DNA onto the center of the dish.
6. Leave the dish in the hood, and gently redistribute the solution from the edges of the dish to the entire surface of the dish after about 10 min.
7. After a total of 20 min exposure to the DNA, add back medium and return the cells to the incubator.

8. After 4 h, remove the medium (aspirate well) and gently add 2.5 mL of HBS containing 15% glycerol (at room temperature).
9. Return the dish to the incubator for 3.5 min. (Do not leave in too long). This time can vary for different packaging lines.
10. Quickly remove glycerol-HBS and rinse (gently) with 10 mL DME. (Dulbecco's Modified Eagle's Medium, GIBCO)
11. Repeat DME rinse and add back 5 mL DME + 10% serum.
12. 18–24 h later, remove the medium and filter through a 0.45 µm filter (the type that easily fits onto a syringe). You can store this "transient" harvest at –80°C or use it immediately for an infection.
13. Add medium (10 mL) back to cells, and continue to cultivate for 1–2 d.
14. If you wish to make stably transfected producer clones, split the transfected cells $1 \rightarrow 20$ or $1 \rightarrow 40$ into medium containing the selective drugs.
15. After 3 d, change the medium to fresh selective medium.
16. After a total of 7–10 days, you should see colonies. Pick well-isolated colonies after 10–14 d.
17. Assay the colonies for virus titer, or use another assay to determine the expression level.
18. When a good producer clone is identified, freeze down many vials. This is to guard against the typical cell culture hazards, as well as the problem of recombination to generate helper virus and loss of titer (some clones reduce their virus output over time, for unknown reasons).

HBS (Hepes-buffered saline)
137 mM NaCl
5 mM KCl
0.7 mM Na$_2$HPO$_4$
6 mM dextrose
21 mM Hepes, pH 7.05

Check the pH. It is a critical parameter.

19. After a producer clone is chosen, some assay for the integrity of the viral genome should be performed on cells infected with virus from the chosen producer clone (e.g., Southern or Northern blots).

3.2.3.2. RETROVIRUS INFECTION AND TITRATION PROTOCOL

1. Preparation of virus inoculum
   A. Plate producer cells (ψ2, ψam, PA317, and so on) so that they become confluent within a week of being plated.
   B. When producers are *really* packed in (usually 1 d past when most people call them confluent), remove the media and discard it. Replace with fresh medium. (Individual producer clones may vary in their ability to produce virus after they become confluent. If this protocol does not yield high titer for your clone, test the producer for virus production by varying cell density, volume of harvest, and time of harvest.)
   C. After 1–3 d, harvest supernatant. It may look quite yellow, and the cells may look sick, but we usually get higher titers if we let the cells incubate for 3 d.
   D. Filter the supernatant through .45 μm sterile disposable filters (*not* .22 μm), and store at –80°C, or use it fresh. We find that we can freeze-thaw stocks at least four times without losing much, if any, titer. You can also add polybrene to the stock and store it with the polybrene already in it, at final polybrene concentration of 8 μg/mL (*see below*).
2. Infection and virus titration using drug selection
   A. Plate cells to be infected the day before they are to be infected. If you are doing a titration, plate NIH3T3 cells 1 → 10 into 60-mm or 100-mm dishes. For other cells, plate them so that they have room to divide at least twice before they become confluent.
   B. Remove medium from cells and add virus stock. We use 1 to 2 mL virus stock to infect a 60-mm dish of cells and 3–5 mL for a 100-mm dish. Include polybrene (Sigma or Aldrich) at a final concentra-

tion of 8 μg/mL. Polybrene stock is made up to 800 μg/mL in distilled H_2O, filter sterilized, and stored at −20°C. Leave the cells at 37°C for 1–3 h (some cells can stay overnight in the high polybrene; other cells are bothered by this. Since the virus only has a half-life of 4 h, longer incubations do not usually result in more infection. Virus absorption takes place fairly rapidly).

C. Add medium back to dilute the polybrene to 2 μg/mL. Incubate for at least two or three cell generation times (2–3 d for NIH3T3 cells).
D. Split infected cells into selection conditions. If the resistance gene is G418, split the NIH3T3 cells 1 → 20 into 1 mg/mL G418 (Geneticin from Gibco) in DME + 10% calf serum, onto 60-mm or 100-mm dishes.
E. After 3 d, change the media. Include drug(s).
F. After 7–10 d (for NIH3T3 under G418 selection), colonies should be visible. Count them, pick them, or both before they spread (usually before 12 d under selection).
G. To calculate the titer:

$$\frac{\text{\# of colonies} \times \text{½ cell split factor}}{\text{vol virus (mL)}} = \text{CFU/mL}$$

Example: Virus (3 μL) was diluted into medium (3 mL DME + 10% calf serum) and inoculated onto a 100-mm dish of NIH3T3 cells split 1 → 10 the night before. Ultimately, 20 colonies were observed on each dish that resulted from the 1 → 20 split into G418. The titer would be 6.6×10^4 CFU/mL. (Multiply by ½ the split factor, since each viral particle could yield several infected daughter cells from the originally infected cell. This is a fudge factor that is meant to reduce inflation, which at its worst would be a factor of 4, if the originally infected cell divided twice after proviral integration. We only use a factor of 2, since we presume that many cells do not yield four daughters prior to G418 selection, but only

yield two daughters.) The volume of virus, rather than the virus dilution, is used in our calculations. This is based on a comparision of titer when 1 µL of virus was inoculated onto 6-cm dishes of cells in the presence of 1 mL, 2 mL, 3 mL, or 4 mL of DME + 10% CS + 8 µg/mL polybrene. Approximately the same number of colonies resulted from each dilution, and thus, we conclude that the amount of virus, not the dilution, is the most important variable (within the limits of the volumes likely to be used in such experiments).
3. Infection and virus titration using x-gal
   A. Proceed through step 2C as described above.
   B. Fix and strain cells as described in 4.1.1. (Do not split cells.)
   C. Calculate titer as follows:

$$\frac{\text{\# of X-gal}^+ \text{ colonies}}{\text{vol virus (mL)}} = \text{CFU/mL}$$

[There is no need for a "split factor." This method is a direct method that should yield the same titer as G418-resistant CFU/mL when comparing titers for a vector such as BAG, which encodes both lacZ and neo (Price et al., 1987)].

3.2.3.3. VIRUS CONCENTRATION
1. Make a virus stock as in 3.2.3.2.
2. Spin virus stock at 14,000 RPM for 20 min at 4°C (J14 rotor in a Beckman centrifuge in 250-mL bottles or in an SW27 rotor for smaller vol). This is to pellet cells and debris.
3. Pour off supernatant directly into fresh bottles, put back into same rotor, and spin 3–16h at 4°C in same centrifuge. (The shorter time may yield a pellet that is likely to break apart, but is easier to resuspend).
4. Discard supernatant, and resuspend pellet gently in 1% original vol. Use DME + 10% calf serum or buffer of your choice. We have not tested a lot of buffers, so many different ones may work. The resuspension may take 2 h. We leave the bottle in an ice bucket in the hood and pipet every 15 min or so. All of the clumps may not dissolve.

5. Store the virus at –80°C. It can be freeze-thawed several times with no loss of titer. Sanes et al. (1986) resuspend it in 50% serum and store for short period of time at 4°C.

3.2.3.4. COCULTIVATION METHOD OF INFECTION
1. Plate virus producer cells (e.g., ψ2-BAG, CRIP-myc, and so on), and allow them to become nearly confluent, or confluent. The degree of confluency should be determined empirically for the particular target cells. (Confluent producer cells may acidify the media to a degree that is harmful for the target cells, but confluent ψ2 producer cells [and perhaps other types of packaging lines] make more virus.)
2. It is often desirable to inhibit replication of producer cells so that cultures of infected target cells will not be contaminated with producer cells or virus after the initial cocultivation (e.g., for such applications as the immortalization of target cells). Cell division of producer cells can be blocked, while allowing for virus production, using irradiation or drugs. For example, gamma irradiation using 2800 rads (1 rad = 0.01 Gy) can be used, or producer cells can be treated with mitomycin C. Producer cells are incubated with 10 μg/mL mitomycin C for 3 h in their usual medium, rinsed several times with medium, and then used for cocultivation.
3. Target cells are plated onto the producer cells and incubated for several days in the presence of 2 μg/mL polybrene. When choosing medium for this step, it may be necessary to make a compromise such that the cocultivation medium is adequate to support cell division of the target cells, while allowing for virus production by producer cells. Allow at least two target cell cycles to elapse during the cocultivation period.
4. If the target cells are nonadherent, they can be washed off the producer monolayer and subcultivated in selective conditions (e.g., G418). Alternatively, the entire culture can be subjected to drug selection by simply replacing the medium with selection medium, as long as the entire culture is not con-

fluent. If the culture is confluent, subcultivation (e.g., split the cells 1 → 10) and drug selection should be performed as described in 3.2.3.2.

3.2.4. Recombination and Helper Contamination

A word of caution is included here concerning the production of wild-type helper virus by packaging lines (Mann et al., 1983; Sorge et al., 1984; Miller and Buttimore, 1986; Danos and Mulligan, 1988; Markowitz et al., 1988). Although the genome that supplies the gag, pol, and env genes in packaging lines does not encode the ψ sequence, it can still be packaged, although at a low frequency. If it is encapsidated with a vector genome, it can undergo recombination in the next cycle of reverse transcription. If the recombination allows the helper genome to acquire the ψ sequence from the vector genome, a recombinant that is capable of autonomous replication is the result. This recombinant can spread through the entire culture. Once this occurs, it is best to discard the producer clone, since there is no convenient way to "cure" the clone of the wild-type virus. As would be expected, this happens with a greater frequency in high titer stocks. The D17–C3, CRE, CRIP, and GP+E-86 packaging lines have a safer design than ψ2, ψam, and PA12. They require multiple recombination events to generate wild-type helper virus. Recombination in these "safe" lines has not yet been observed and is predicted to be of an extremely low frequency. It is possible to monitor producers for the production of wild type. A protocol that is very sensitive for this purpose is included (3.2.4.1).

3.2.4.1. HELPER ASSAY. This is the assay that we run to detect helper virus. There are other assays [reverse transcriptase (Goff et al., 1981), and so on], but this one is very sensitive and easy to run. The example below is for a neo virus-producing line. If you are using a different drug marker, substitute that selection for G418. Also, if you wish to test the host range of the helper, you can assay the final supernatant (in step 6. below) by infecting both NIH3T3 cells and a nonrodent line, such as the dog line, D17. This assay is designed to detect the production of neo virus by cells that are infected with "helper-free" stocks. Cells infected with helper-free virus should not be able to produce neo virus at high titer, although some passive carry-over of defective helper genomes may lead to low titers (Danos and Mulligan, 1988). However, cells that are infected with a helper-contaminated supernatant will produce a neo virus at high titer, which is detected by harvesting and assaying the supernatant on uninfected cells.

1. Make the virus supernatant to be tested as potent as possible. We usually harvest the supernatant from confluent cells that have been confluent for 2 to 3 d and that had the medium replaced with fresh media on the day that they reached confluence (as in 3.2.3.2).
2. Split NIH3T3 cells the day before, 1 → 50, and plate onto a 60-mm dish. Set up one dish for a positive control, one dish for a negative control, and one dish for each stock to be tested.
3. Infect the NIH3T3 cells with 1 mL of the virus stock (from step 1.) using 8 µg/mL polybrene, as usual (protocol 3.2.3.2). Also, make sure that you filter the virus stock through a .45-µm filter before doing the infection. For a positive control, coinfect a dish with a known helper-free neo virus and wild-type helper (i.e., virus produced by a wild-type producer cell line). For a negative control, use a known helper-free neo virus. After allowing 1–3 h for absorption, add 4 vol of media. Make sure that you keep the polybrene at 2 µg/mL for the length of the assay to allow for horizontal spread (i.e., infection of neighboring cells). (As an optional, additional negative control, one dish should receive no virus, just polybrene and media.)
4. After the cells become confluent (3–4 d), split 1 → 50 into 60-mm dishes, 1 dish/assay. Again use 2 µg/mL polybrene.
5. The day that the infected cells become confluent, split fresh, uninfected NIH3T3 cells 1 → 20, to be infected the next day. (Again, use 1 dish/assay.) Also, throw out the old media on the confluent (putative producer NIH3T3) cells, and replace it with ½ vol of fresh DME + 10% calf serum.
6. Harvest the supernatant from the confluent cells, filter through .45-µm filter, add 8 µg/mL polybrene, and infect the cells that you split yesterday with 1 mL of the supernatant.

7. After 2–3 d, split the newly infected cells 1 → 20 into media with 1 mg/mL G418, just as you would normally do for a neo virus assay (protocol 3.2.3.2.). Change media after 3 d, and count colonies after 10 d.

If there is passive transfer of the ψ-minus helper genome, there will be a low titer of neo virus. If there is an intermediate titer, there may be a "crippled" helper because of a partial or inaccurate recovery of the ψ sequence by the helper genome during recombination. If there is high titer (e.g., as high as the original titer in the stock produced by the packaging line), then a wild-type, or nearly wild-type virus, is present. To distinguish between passive transfer and crippled helper, the inoculum used in step 6 can be used for an additional run of the procedure, beginning at step 2. If there is passive transfer, the low titer stock will not continue to yield virus upon subsequent passage. A crippled helper can continuously make low to moderate titer.

3.3. Gene Expression

The viral promoter (LTR) has been used in many vector designs to express a wide variety of genes taken from many sources. Genes from other viruses (bacteria, yeast, birds, and mammals) have been successfully transduced and expressed. Genomic DNAs as well as cDNAs have been used. Basically, any gene is likely to be transducible, but one can't predict the level at which any given one will allow viral transmission and expression. Some LTR-promoted genes result in high levels of expression at the RNA level, but give low levels of protein expression. The reason(s) for this is unclear. The LTR promoter appears to be fairly nonspecific in terms of expression in a wide variety of cell types. However, there is evidence that it is inactive in preimplantation mouse embryos and early stem cell lines (Jaenisch and Berns, 1977; Teich et al., 1977; Gorman et al., 1985). There have also been problems in expression in bone marrow cells, the cells that were the target for the first gene therapy experiments in humans (Williams et al., 1986; Magli et al., 1987). Intense efforts at alleviating the expression problems in this system may provide useful insights into expression in the nervous system should similar problems arise.

Promoters other than the LTR have been used successfully in several types of vector designs. Internal promoters that direct transcription in the same direction as the LTR have been quite popular (for example, see DO-L in Korman et al., 1987). These internal promoters can be nontissue specific, or can be used to direct transcription in a specified tissue (e.g., the β-globin promoter, Cone et al., 1987b). Promoters can also be placed in the opposite transcriptional orientation, relative to the LTR. In this case, one must include a polyadenylation site and perhaps splicing signals. Some of the internal promoter vectors include LTR activity, whereas others do not. Deletion of the viral promoter and/or enhancer activity may be desirable if one wants to eliminate any potential activation or suppression of an internal promoter via the enhancer or promoter of the LTR.

For the application of these vectors to the problem of lineage mapping (discussed in 4.), a promoter that does not exhibit cell-type specificity is required. If a promoter were used that only expressed in a few cell types, it would select the spectrum of lineages that could be mapped, and could even limit the results to the most obvious or trivial of findings. However, the choice of such a constitutive promoter is not a simple one. Promoters for housekeeping functions seem an obvious starting point. However, when one is hoping to map lineages of cells that are postmitotic, and as specialized as neurons, not all housekeeping promoters would be expected to be expressed as fully as they might be in mitotic cells. Furthermore, the specificity of a promoter when it is inserted into the context of a retrovirus is currently a matter of investigation. One must perform experiments with all of the proper controls to establish the efficacy and specificity of the internal, non-LTR promoter. The best cases for the retention of specificity of internal promoters in retrovirus vectors are the human β-globin (Soriano et al., 1986; Cone et al., 1987b; Karlsson et al., 1987) and the light chain immunoglobulin (Cone et al., 1987a) promoters.

Expression in brain tissue of the wild-type MMLV LTR was demonstrated by infection of postimplantation mouse embryos (Jaenisch, 1980). Since no histochemistry was performed in these studies, the identity of the neural cells that were expressing the viral RNA was not resolved. Sanes et al. (1986), Calof and Jessell (1986), and Price et al. (1987) have demonstrated the ability of the MMLV LTR promoter, as well as the SV40 early promoter transduced by a retrovirus vector, to express in neural tissue. Neuroblastoma and glioma cell lines are infectable, as are all such cell lines that have been tested. In addition, primary cultures of cerebral cortex, olfactory bulb, retina, cerebellum, neural crest (rat and mouse), and dorsal root ganglion are infectable. Thus, we constructed the BAG virus using the viral LTR for our first test virus for the lineage mapping protocol (Price et al., 1987); Sanes et al. (1986) used the SV40 early promoter; Calof and Jessell (1986) also used the LTR. It remains to be documented whether these promoters will function in all cell types in vivo. Caution is extended here, since expression in vitro does not necessarily imply that it will occur in vivo as well.

4. Applications to Neurobiology

The applications of retrovirus technology to the study of the nervous system are virtually the same as for any other system. Lineage mapping and immortalization of neural cells are the only topics discussed here because of space limitations. However, a typical application of retrovirus vectors is in the study of expression of a gene for examination of its activity. The reduction of a genomic DNA to a cDNA (Shihmotohno & Temin 1982; Cepko et al., 1984) is also a useful application for neurobiologists that has not yet been exploited.

4.1. Lineage Mapping

Lineage mapping via the introduction of a traceable marker is a classical technique in developmental biology. A variety of markers and techniques exist, but most are short-term and have been limited to large, accessible cells. Retrovirus vectors provide an opportunity to deliver a stable, genetic marker of one's choice. The application of this technique to the analysis of lineage in rat retina (Turner and Cepko, 1987), mouse epithelium (Sanes et al., 1987b), chick tectum (Gray et al., 1988), and rodent cortex (Walsh and Cepko, 1988; Luskin et al., 1988; Price and Thurlow, 1988) has been quite successful, and suggests the general utility of this approach. The parameters of this application will be discussed below.

4.1.1. Histochemical Markers

The choice of a gene for use in lineage mapping is based on several criteria. First, it is quite useful to have a gene product that enables one to use a known, enzymatic, histochemical stain for easy identification of infected cells. Enzymatic histochemical staining is not an absolute requirement, as immunohistochemical procedures are quite standard, but enzymatic reactions can be easier and more sensitive. They also provide flexibility for secondary characterization of marked cells with antibodies that serve to identify cell types. For example, if one encounters a clone of cells that are marked with the lineage tracer (an enzymatic stain), one can then use a rabbit antineurofilament and a mouse antiglial filament antibody stain (coupled with rhodamine and fluorescein conjugated second antibodies) to identify the nature of the infected cells. A second set of criteria concerns the gene itself. It is helpful if it is already cloned and characterized with respect to its molecular structure. It should also be innocuous, have no effect on development, and not be tissue-specific. One must be able to detect it in tissue that already contains many enzymatic activities. Either overexpression of the introduced gene product, relative to the endogenous activity, or a unique gene product or specific assay method is required. Finally, the gene must be transmissable in a retrovirus vector. Although it is possible to transmit more than one (possibly up to three) gene in a retrovirus vector, a single gene is much easier to work with. In general, genes from any source can be used. One must try to match the vector and gene, and empirically determine whether high titer transmission and high level of protein expression are possible.

The $E.$ $coli$ lacZ gene encoding β-galactosidase (β-gal) has been used in tissue culture cells in vitro (Hall et al., 1983; Nielson et al., 1983; Norton and Coffin, 1985), $Drosophila$ in vivo (Lis et al., 1983; Hiromi et al., 1985) in transgenic mice (Goring et al., 1987) and in many bacterial strains. The β-gal enzyme yields an intense indigo precipitate after hydrolysis of the substrate, X-gal, thus enabling identification of individual infected cells (Pearson et al., 1963; Lojda 1970). We have found that β-gal is expressed at high levels in infected cells after infection with the BAG virus (Price et al., 1987; Turner and Cepko, 1987). Sanes et al. (1986) similarly found good expression from the SV40 early promoter. Other genes that have been used to mark cells in this manner include luciferase (a firefly gene used to mark plants; Ow et al., 1986) and *Drosophila* alcohol dehydrogenase. This latter enzyme has been used as a histochemical stain in *Drosophila* (Ursprung et al., 1970) and was recently employed to examine lineage in early mouse embryo development by Roger Pederson and his colleagues at UCSF. It may prove to be a generally useful marker. Protocols for X-gal staining of cells in culture and tissue are given below (4.1.1.1. and 4.1.1.2.).

4.1.1.1. XGAL STAINING OF CULTURED CELLS. We currently stain cultured cells with X-gal using either of two protocols. The choice of protocols is based primarily on convenience, as both work quite well, although the glutaraldehyde fix gives slightly bluer cells.

Protocol #1

1. Fix cells in PBS with 0.5% glutaraldehyde for 5–15 min at room temperature. We buy 25% glutaraldehyde from Sigma and freeze-thaw it many times. If you wish to stain some wells in a macro or microtiter dish, care must be taken to prevent the fumes from killing the cells that you don't want to stain. Prevent this killing by leaving the lid off the dish and working in the hood, until the actual staining step, when the lid is placed back and the dish is returned to the incubator.
2. Rinse the cells well 3× with PBS (137 mM NaCl, 2.7 mM KCl, 8 mM Na$_2$HPO$_4$, 2.6 mM KH$_2$PO$_4$). Leave the second rinse on for about 10 min, and perform washes 1 and 3 quickly. Residual fix will inhibit the enzyme reaction.
3. Add staining solution. Use a minimal vol to cover the cells, since X-gal is very expensive.
4. Incubate at 37°C for 2 h to overnight.
5. You can store the dish at 4°C indefinitely.

Protocol #2

1. Fix the cells in 2–4% paraformaldehyde in PBS. Use a paraformaldehyde solution that is well buffered and less than a week old. You can leave it on for up to 60 min without harming the enzyme activity.
2. Wash and proceed as above.

Xgal Mixer [modification of recipe from Lojda, 1970 and Dannenberg et al. (1981)]:

5–35 mM K$_3$Fe(CN)$_6$
5–35 mM K$_4$Fe(CN)$_6$·3H$_2$O
1–2 mM MgCl$_2$ or MgSO$_4$
In phosphate-buffered saline (PBS)
This solution can be stored at least a few weeks at room temperature.

When ready to use, add the X-gal (5-bromo-4-chloro-3-indolyl β-D-glactopyranoside) to a final concentration of 1 mg/mL. Make up the X-gal at 40× in dimethylformamide, and store in glass covered with foil at –20°C.

Notes on Protocols 1 and 2

If you are staining cells as part of a titration of a lacZ containing (3.2.3.2.) virus, wait at least 48 h after infecting before staining. We checked to see the minimal time necessary for most newly infected cells to score as blue, and it was 48 h. If you wait longer, it will not really matter, except that clones may be larger and harder to recognize as clones if the cells migrate extensively.

Background staining is not usually a problem in the CNS except with activated macrophages. If you have a problem, shorten the incubation time. We can usually recognize infected cells within 30 min.

The amount of ferric and ferrous cyanide to use depends on your taste. We think that the higher amount causes the precipitation of the indole to occur more quickly and thus reduces diffusion. However, it may cause a greenish background upon prolonged incubation (overnight or longer) in some tissues.

4.1.1.2. X-GAL HISTOCHEMISTRY FOR TISSUE
1. General procedure for any tissue
   A. Make up a fresh paraformaldehyde solution as described below, or use 0.5% glutaraldehyde in PBS. (Glutaraldehyde may give background in some tissues.)
   B. Either dissect out the tissue of interest (if it is quite small, *see* whole-mount procedure below) or perfuse animal with fix. We perfuse adult mice for about 30 min, let the animal sit for another 30 min, and then dissect (if appropriate).

C. The tissue is next incubated at 4°C in PBS + 2 mM MgCl$_2$ + 30% sucrose until the tissue sinks (a few hours or overnight).
D. The tissue is embedded in OCT compound (Miles) and frozen on dry ice.
E. Cryostat sections (20 μm to 100 μM) are cut and placed on gelatin-coated slides. The sections are air-dried. They can be stored at 4°C.
F. Have ready cold PBS + 2 mM MgCl$_2$ and PBS + 2 mM MgCl$_2$ + 0.01% sodium desoxycholate + 0.02% NP40. Fix the sections again in cold fix (below) for 5 min. Rinse quickly once in cold PBS + MgCl$_2$, then again in the same buffer for 10 min, then permeabilize by incubating in the PBS + detergents for 10 min, also at 4°C.
G. Stain in X-gal solution (see 4.1.1.1.) for a few hours, up to overnight, at 37°C.
H. Rinse 3× in PBS (for a few minutes each rinse) and mount in gelvatol (Air Products and Chemicals, Inc. Allentown, Pa., see Rodriguez and Deinhardt, 1960).
2. Whole-mount procedure

If you have a small piece of tissue, such as a retina or early embryonic tissue, you can process it as a whole mount. Fix the tissue for 5–15 min, as described above, and wash the tissue in PBS + 1 mM MgCl$_2$ 3× for 5 min each. Incubate in the X-gal staining solution, with or without detergents. In retinal whole mounts, the detergents are not necessary.

Paraformaldehyde solution

Dissolve solid paraformaldehyde (2 g) (although supplier may not be important, we use BDH Chemicals, LTD. Poole, England) in 100 mL of .1M pipes buffer (30.24 g/L, pH = 6.9) containing 2 mM MgCl$_2$ (200 μL of a 1 M stock) and 1.25 mM EGTA (250 μL of a .5M EGTA stock at pH = 8.0). Heat in the hood with stirring for about 5–10 min. Don't boil. Cool to 4°C.

4.1.2. Methods of Delivery

Retroviruses enter cells via interactions at the host cell surface. Thus, it is not necessary to perform intracellular injections to mark cells. Indeed, it is not known if direct injection of a virus would permit infection to occur. Placement of viral inoculum in the vicinity of dividing cells is all that is required to initiate infection. The virus has a short half-life (4 h at 37°C, unpublished observations) and absorbs quickly to cells. One can therefore assume that infectious, extracellular virions disappear very soon after virus is delivered. (Virus can also absorb to cells that do not bear receptors, although this is low affinity, nonspecific "sticking." It does not lead to infection.) Virus that enters postmitotic cells probably undergoes reverse transcription, but fails to integrate. Unintegrated viral genomes do not express a detectable level of RNA. If one performs an injection into an area that contains a majority of postmitotic cells, one will find very little, if any, evidence of viral infection. In general, one is limited to a virus titer of 10^6–10^8 CFU/mL, and to a small vol (.1–1 μL). It is therefore quite important to make the best attempt possible to deliver virus directly to the mitotic zone. It is unclear if there are factors in tissue fluids (e.g., ventricular fluids or CSF) that inhibit or destroy viral infectivity.

Virus delivery to postnatal animals is fairly straightforward. We use a hand-held Hamilton syringe with a 33-gage needle to deliver virus to the mitotic zone of the retina (Turner and Cepko, 1987). Alternatively, we infect postnatal mouse cerebellum and olfactory bulb using pulled glass pipets. The skull is soft enough the first few days after birth for direct injection into the tissue through the skin and skull. Coinjection with a dye, such as 0.05% trypan blue or fast green, aids in the ability to detect the accuracy of injections and does not impair virus infectivity. The animals are anesthetized by cooling at –20°F in a freezer or on ice for a few minutes. The animal is then illuminated using a fiberoptic light source, and the solution is injected directly into the desired area using a hand-held pipet. It is best to practice a series of injections with dye alone and then immediately dissect the animal for examination of the injection. It is also valuable to inject cells that contain the β-gal gene, such as the ψ2 BAG producer cells (available from ATCC #CRL 9560). This is not only a good way to check the accuracy of injection, but it is also a good control for the X-gal histochemistry technique. The cells should be well trypsinized, pelleted, and washed twice with DME or PBS, and then resuspended to a concentration of 10^8 cells/mL. If one injects 0.1 μL successfully, the cells can easily be found using the X-gal staining technique. The animal should be killed a few minutes after the injection and processed for X-gal histochemistry. An internal control of your X-gal histochemistry can be accomplished at the same time. Prior to injection, the cells can be prelabeled with a fluorescent dye, such as the carboxyfluorescein diacetate succinimyl ester of Molecular Probes Inc., Junction City, Oregon. As detailed by Bronner-Fraser (1985), the cells are exposed to 0.3 mM CFSE at 37°C for 30 min (make up a 10-mM stock in DMSO and store in foil in the refrigerator) in PBS just before trypsinizing. The injected cells can then be monitored independently from the X-gal staining by viewing the slides under a fluorescent microscope.

If early lineage data in the nervous system is desired, virus must be delivered to embryos. Embryos can then develop *in utero*, *exo utero* (but still within the mother), or in vitro. Jaenisch (1980) and Sanes et al. (1986) performed viral injections on mid-gestation mouse embryos. These injections were undirected, through the uterine wall and deciduum. Both groups were successful, although it should be noted that, in the case of Jaenisch, the virus was replication-competent, and thus even if only one particle initiated infection, a full infection of the embryo could occur through viral spread. Sanes et al. were able to reproducibly mark cells on the outside of the embryo and in the yolk sac. They were able to identify lineages in the visceral yolk sac and skin. Luskin et al. (1988), using E12 and E13 mouse embryos, Walsh and Cepko (1988) and Price and Thurlow (1988), using E14–E19 rat embryos, were also able to inject through the uterine wall into the lateral ventricle. These injections resulted in infection of the appropriate cortical progenitors throughout the cortex.

Soriano and Jaenisch (1986) infected early mouse embryos (4–16 cell stage) in vitro and then reimplanted the embryos into pseudopregnant hosts for full development *in utero*. They were interested in the number of cells that give rise to the entire embryo and the lineage of somatic vs germ line tissue. No histochemistry was used to mark single cells. The lineage marking was through the use of viral integration sites, each of which is essentially unique. Southern blots were used to determine clonal relationships. This option was available because the clones were extremely large since they infected progenitors of the entire mouse. Similarly, in a study of hematopoietic lineage, Lemischka et al. (1986) used integration sites to mark transplants of infected bone marrow. In most of the applications considered here, the clones will contain far less than the 10^4–10^5 cells necessary for examination of viral integration sites, using conventional Southern blot technology. However, experiments can be devised for marking very early progenitors of the nervous system, and thus, integration sites and Southern blot technology could be used as clonal tags. Alternatively, the new and extremely sensitive technique known as polymerase chain reaction, or PCR, may enable the examination of integration sites from single, or very few, cells (Saiki et al., 1985, 1988; Ochman et al., 1988).

Calof and Jessel (1986) employed an alternative approach to marking early embryos. They prepared in vitro cultures of midgestation mouse embryos and infected them by injections into the amnionic cavity. In this case, development can proceed only for a few days. Calof developed an immunohistochemical marker virus by using the human T8 surface marker. Clearly labeled cells suggestive of neural ectoderm and surface epithelium were observed 24 h later. The exact nature of these small clones was difficult to determine. In vitro approaches are useful for certain directed questions about lineage.

Injections that are precisely directed into specific mitotic zones of the early mammalian nervous system greatly benefit from the technique of Muneoka et al. (1986). They were able to manipulate E10–E14 mouse embryos surgically in the mother. They removed the embryo from the uterus, leaving it attached to the placenta, and performed surgery on it while allowing it to float in the abdominal cavity in a buffered saline solution. Remarkably, these embryos can be brought to term in the abdominal cavity and can be delivered by Caesarean section at the end of gestation. If this technique can be generally mastered, it may prove of great value for injections into specific brain regions. In any case, one can perform rather crudely directed injections into the ventricles or into selected areas of the brain on E12–E14 embryos, using the technique of injecting through the uterine wall, as described above. Turner et al. (1988) recently injected the BAG virus into the mitotic zone of the retina of *exoutero* embryos. These were brought to full term, delivered by Caesarean section, and shown to be infected. Well-directed, precise injections should prove more useful in the long run for lineage analysis.

Retroviruses can also be used to infect species that are more accessible during embryogenesis. Avian embryos seem to be prime candidates for this procedure. They are not only accessible, but naturally occurring avian retroviruses have been adapted as vectors (e.g., Hughes et al., 1987; Stoker and Bissell, 1988). Murine xenotropic retroviruses can infect chickens (Levy, 1975), and we demonstrated that the BAG virus encapsidated in the amphotropic murine coat effectively infected chicken cells in vitro using embryonic fibroblasts and retina organ cultures (Price et al., 1987). However, in vivo infections using amphotropic viruses have been extremely inefficient. Recent progress in the development of avian vectors and packaging systems may make gene transfer into avian cells as efficient as in mammalian cells. As mentioned above, Stoker and Bissell (1988) made a stable ALV-based packaging line, Q2bn. Although it is not clear how well this will function to produce high-titer stocks, it is an important step in the right direction. Gray et al. (1988) have made avian vectors that express lacZ in vivo and used their stocks to generate clones in the chick optic tectum. They demonstrated radial alignment of sibling cells in the majority of the clones and should be able to perform lineage analysis in the future. Infection of avian species will certainly be exploited in the near future, although expression from the Rous Sarcoma Virus (an avian retrovirus) LTR in very early chick embryos (stage X–XI) appears to be repressed (Mitrani et al., 1987). Later expression appears to be quite high, but tissue specificity of expression has not been extensively analyzed.

4.1.3. Analysis of Clones

After the introduction of the virus into the mitotic zone, one must allow time for infection to occur, progeny cells to be generated, and in some cases, for migration and/or differentiation. The time of analysis will vary depending upon the system and the information that is desired. The most complete lineage map can be generated if one injects at different times in development and allows a range of survival times before analysis. If migration is a key component to differentiation of the tissue, analysis at early times after infection may be necessary to allow recognition of a clone. On the other hand, if a cell type can only be recognized after allowing differentiation to occur, and if differentiation occurs after migration, very little information may be gained from an early analysis.

Recognition of a clone as such and confidence in one's assignment of clonal identities are key issues in the use of this technique. We have greatly benefited by working on a well-described tissue that is beautifully laminated and that exhibits very little lateral migration (Turner and Cepko, 1987). We label the mitotic zone of postnatal rat retinas, as described above, and wait for 5 d to 12 mon before harvesting the tissue. Since we can prepare the tissue for histochemistry as a whole mount, it is possible to determine if the infection was successful within a few hours of harvesting. The tissue can be sectioned and cells identified on the basis of location and morphology. In the retina, interpretation of labeled cells as clones is quite straightforward, because the migration of the progeny cells away from the mitotic zone occurs in a strictly radial fashion. Since the amount of virus that we can inject is limited by both the volume and titer, the frequency of infection is only about 1 in 100 to 1 in 1000 of the mitotic cells at the time of infection. Thus, clones are generally spaced well apart, and it is usually not difficult to recognize clonal relationships. However, the real basis for concluding that each cluster of labeled cells is indeed a clone is from infections with diluted virus stocks. In a set of injections with 10-fold serial dilutions of a virus stock, there is a linear decrease in the number of clones that are obtained. Moreover, the composition and size of the clones remains the same, regardless of the viral dilution. This is the classical argument for a single hit event used by virologists to determine the number of infectious units required to effect a given event.

The recognition of clones within a tissue that undergoes a great deal of migration in different directions is proving to be difficult (Walsh and Cepko, 1988; Price and Thurlow, 1988). The use of mixtures of different marker viruses may be one way to solve this problem. If these were available, one could coinject two or three viruses that yield distinguishable colors or markings. Alternatively, very dilute infections where only a minority of animals exhibit any infected cells could allow for statistical arguments of clonality, and the PCR analysis mentioned above could be developed to provide a biochemical mark of clonality.

Infection of tissue explants followed by in vitro cultivation or transplantation into a host animal is another application of the retrovirus marking system. For example, we used in vitro culture of early chick retinas to examine retinal lineage and also marked transplanted retinas by infection of embryonic rat and mouse retinas, which were then transferred to postnatal animals in ectopic locations [using the protocol of Hankin and Lund (1987) (Cepko and Lund, unpublished)]. Emson et al. (1988) infected rat fetal striatal neurons prior to transplantation into lesioned adult rat striatum. They were able to recover βgal⁺ cells 8 mon later from the striatum. The BAG retrovirus was similarly used by Short et al. (1988) to mark transplanted cells that were grafted into the CNS of newborn and adult rats. Price et al. (1987) and Luskin et al. (1988) marked in vitro cultures of cortical cells to investigate lineage. The use of cells cultured in vitro should allow manipulation of conditions in order to determine directly the factors that influence lineage decisions.

If a reliable vital marker were developed, it would greatly enhance the power of retroviral marking by allowing for a dynamic view of marked cells located in vivo or in vitro. Although a vital, fluorescent stain for βgal is now available (fluorescein di-β-D-galactopyranoside, Molecular Probes, Eugene, OR), the fluorescent product diffuses out of positive cells too rapidly to permit visualization of positive cells in sections or in whole mounts (unpublished observations). However, this substrate can be used to sort positive cells on a fluorescence-activated cell sorter, a very useful feature for some applications (Nolan et al 1988).

4.2. Immortalization of Neural Cells

Retroviruses were first recognized by their ability to transform cells and induce tumors in animals. The acute transforming retroviruses accomplish this by transduction and expression of oncogenes (for reviews *see* Bishop, 1985 and Adamson, 1987). Some of these same oncogenes, as well as others, have been recognized as potentiators of oncogenesis independently from their transduction by retroviruses. Oncogenes are so named, since they induce immortalization or transformation in cells in vitro and/or in vivo. For the purposes of this discussion, these two phenomena will be considered in simple, operational terms. Immortalization is the continual, indefinite proliferation of primary cells in vitro. Transformation is the acquisition of altered growth properties that endow the cells with the ability to form foci in vitro and tumors in vivo. Various oncogenes induce one or both of these properties (Land et al., 1983). Oncogenes have also been categorized based on their intracellular location and possible mechanism of action. Many oncogenes appear in the nucleus and are hypothesized to play a role, either directly or indirectly, in gene regulation.

Others exhibit tyrosine kinase activity and are localized to the cytoplasm, sometimes in association with the plasma membrane. Growth factors and their receptors can also act as oncogenes.

Transduction of oncogenes via gene transfer techniques has led to establishment of lines representing a variety of cell types (e.g., hematopoetic, muscle, fibroblast, melanoblast). In some cases, these lines have greatly aided in understanding aspects of development (e.g., rearrangement of immunoglobulin genes in Abelson-transformed B cells; Alt et al., 1981). The aim in this section is to provide guidelines and protocols for the establishment of neural cell lines via oncogene transduction in retroviruses. Reviews of this topic have recently been published (Cepko, 1988, 1989).

4.2.1. Choice of Target Cells

4.2.1.1. LIMITATION TO MITOTIC CELLS. As retroviruses require that the host cells undergo an S phase for viral integration, mitotic cells must be used as targets for immortalization. Postmitotic cells, such as neurons, do not integrate or stably express retroviruses such as MMLV (the retrovirus used as the parent of most vectors). However, there is a murine neurotropic retrovirus that appears to infect postmitotic motor neurons in vivo (Gardner, 1978). More work needs to be done to document this finding. It would be informative if experiments were performed on postmitotic cells in vitro. If this virus does indeed possess the ability to infect neurons, it would be a useful vector for some applications. Recently, a vector from Herpes simplex virus (HSV) was shown to transduce and stably express lacZ under control of an HSV immediate early promoter in postmitotic peripheral neurons (Geller and Breakefield, 1988). Direct transfection of postmitotic cultured neurons was also recently shown to result in expression of transfected genes (Werner et al., 1988). If oncogenes could be stably associated with postmitotic neurons via either of these methods, it might be possible to immortalize differentiated, postmitotic cells. However, even if one could stably associate oncogenes with postmitotic cells, it remains to be demonstrated whether it is possible to maintain the mature, differentiated neuronal phenotype in cells if they are induced to undergo cell division. Immortalization of progenitor cells, or postmitotic cells, followed by arrest of oncogene expression and/or application of conditions that promote differentiation into mature neural cells, may be the best strategy for the production of lines with desirable properties, regardless of whether the oncogene is introduced via retrovirus, HSV, or transfection.

There are temperature sensitive (ts) alleles of several oncogenes. The ts alleles of src (Wyke, 1973; Giotta et al., 1980) and SV40 large T antigen (Tegtmeyer, 1975; Fredericksen et al., 1988) have been used for immortalization of neural cells. Mitotic cells immortalized by these alleles can be maintained at the permissive temperature (PT) of 32°C, and differentiation can be induced at the nonpermissive temperature (NPT) of 39°C. There are also ts alleles of myc (Weizacker et al., 1986) and polyoma large T antigen (Eckhart, 1969). The full extent of differentiation of cells at the NPT still remains to be investigated or documented. Alternatives to ts genes for induction of differentiation include growth conditions in vivo and in vitro, and the use of inducible promoters. To date, an inducible promoter that is truly inactive under noninduced conditions has not been available in the context of a retrovirus, but it should be possible to construct such vectors. Growth conditions, including cocultivation with primary cells alone or in conjunction with temperature shift (when a ts oncogene is used) may prove the most useful for full induction of differentiation.

4.2.1.2. SELECTION OF TARGET CELL. If one has a specific cell type that is the target for immortalization, it may be possible to select in some way for that cell prior to the infection. (Screening and selection techniques for cells after immortalization will be discussed below). At a minimum, if retroviruses are used, it is necessary to know the best source of tissue for the mitotic progenitor. It is then extremely valuable to also know something of the antigenic or biochemical characteristics of the progenitor. The problem that is often encountered in this approach is recognition of the immortalized cells as the mitotic progenitors of defined, differentiated cells. This lack of knowledge concerning the progenitors is a major hurdle for progress in this area. If there is a feature of the progenitor that can be identified, it may be possible to sort the progenitors, or perhaps directly select them, prior to the infection. The best reagents are monoclonal antibodies or ligands that recognize surface molecules. Using the fluorescence-activated cell sorter, one can directly sort the cells. Alternatively, one can coat a dish with the antibody and select by the technique of differential adhesion, or "panning" (Aruffo and Seed, 1987). Use of a cell type-specific promoter may also allow immortalization of specific target cells (Largent et al., 1988). The ability to select specific target cells will greatly reduce the amount of work encountered at later stages.

4.2.2. Choice of Oncogene and Vector

The best oncogene for a given application presently can only be determined empirically. Our rationale has been to use several oncogenes of the immortalizing type (Ryder et al., 1987, 1988; Hen et al., 1987; Roberts et al., 1985; Jat et al., 1986; Land et al., 1986). Our hope has been that the cells will retain more of their nontumorigenic properties and be more likely to differentiate than they would if they were transduced with a transforming oncogene. However, no general statements can be made and almost any choice can be rationalized. It is best to try several different oncogenes and express them at different levels. The choice of vector is also empirical. As mentioned above, it appears that vectors in ecotropic capsids gain entry into most cell types. They can integrate and express from the viral LTR in most mitotic cells (the exceptions, as noted above, appear to be preimplantation embryos and cell lines that represent this tissue). There are thus many vectors that appear capable of transducing and expressing the oncogene of choice. Indeed, many such constructs have been made or occur naturally. The "best" choice of the oncogene and vector may then be the constructs that are available to you. If the target cells are nonrodent, then the number of constructs and the hazards associated with their use are much more limiting. If one wishes to immortalize human cells, the vectors are by definition capable of infecting the experimenter, and great care must be exercised. It is best in this case to choose the CRIP helper line since it has not yet given rise to helper virus. Since the full host range of cell types for infection by the amphotropic virus has not been investigated as thoroughly as the ecotropic virus ($\psi 2$ virus), it is not clear how effectively one can deliver vectors to nonrodent neural progenitor cells. However, it is likely that the amphotropic-coated vectors will have a broad host range.

4.2.3. Experimental Strategies for Production of Cell Lines

4.2.3.1. INFECTION OF PRIMARY CELLS. A straightforward strategy for transduction of oncogenes is to make primary, dissociated cell cultures and infect them using the protocols for other cell types (Protocol 3.2.3.2 or 3.2.3.4.). One starts with tissue containing the mitotic progenitor and hopefully has some idea of the best conditions for proliferation in vitro. This should not be confused with conditions that may be described for maintenance or growth of the differentiated cells of interest. For example, serum-free medium devised for physiological studies of a particular neuronal population may not support mitotic activity of the progenitor. If conditions for the progenitor's growth are not known, it is advisable to try more than one rich medium that could support growth of neural cells. Inclusion of growth factors and use of substrata that promote adhesion (poly-*l*-lysine, laminin, and so on) are recommended. Several groups (Ryder et al., 1987, Frederiksen et al., 1988, and Bartlett et al., 1988) have used DME plus 10% fetal calf serum and poly-*l*-lysine or poly-*l*-ornithine-coated plastic dishes for cells from several sources (olfactory bulb, cerebellum, cerebral cortex, and embryonic mesencephalon).

Dissociation techniques that yield single cells, high mitotic index, and viability are recommended. If the cells remain in large clumps, the virus will have access only to the cells on the outside of the clump as viruses do not possess any invasive properties. If single cells are not optimal for viability, small clumps are still satisfactory, because it is not necessary to infect a high percentage of cells. It is best to opt for high viability and mitotic activity. We typically dissociate using 0.5% trypsin. We then infect using a standard infection protocol (3.2.3.2.) including polybrene, after the culture shows signs of good health. This may be the day following plating, or it may be up to several days later. The medium is removed carefully to avoid removing cells, and virus supernatant is applied. After 1–3 h at 37°C, the viral supernatant is removed and the medium of choice is added back. An alternative method for infection calls for cocultivation of producer cells with target cells (protocol 3.2.3.3.). The intimate contact between producer and target cells may promote a higher efficiency of infection. Producer cells may also provide a feeder layer function for the target cells. Bartlett et al. (1988) used this method for successful infection of embryonic mouse mesencephalon.

Following exposure to virus, at least several days (minimally two cell cycles) are allowed for incorporation and expression of the vector genome regardless of whether infection is via cocultivation or cell-free supernatants. The cells are then placed in medium that selects for the viral genome. Usually, the vector will encode the neo gene, and thus, the selection drug will be G418. The concentration of G418 should be carefully titrated for each system. Primary neural cells are quite sensitive to this drug. The minimal dose that kills all uninfected control cells should be used. We have found this to be about 0.3 mg/mL for most of our rodent CNS cells (and neural crest cells). The infected cells can be exposed to the selective conditions in two ways. If the culture is fairly sparse and slow growing, the drug can be added to the culture medium. Alternatively, if it is a fast growing culture that is fairly dense, the cells should be subcultivated at the time of drug addition. We typically split nearly confluent dishes of cells 1 to 10 or 1 to 20. The number of G418 resistant colonies that results will greatly depend on the titer of the virus and the mitotic index of the cells at the time of infection. For example, if the virus titer is fairly high, but the mitotic index is low, the number of colonies will be low. The virus will not target to the mitotic cells, but will be absorbed by all receptor-bearing cells on the dish. Optimizing for high mitotic index and enriching in some way for the cells of choice are two aspects that are truly worth the effort.

One may ask why drug selection is required at all, since the presence of the oncogene may select the cells carrying the virus. In some cases, this may be so. In other cases, depending on the growth conditions and the culture, there may be glial, fibroblast, meningeal, or endothelial cell growth in the absence of infection. This can obscure or inhibit the growth of other types of cells in the first few weeks in culture. We have found it quite useful to be able to examine the morphology of clones of cells that are well separated from each other. The colonies grow at widely varying rates, and morphology changes can take place within a colony over the period of a week. Without the ability to focus on individual colonies, such changes would not be appreciated. In cases where a preselection was made, and a well-defined population is in culture, the drug selection may not be useful. However, subculturing the cells at the time of drug selection is a good means of cloning a line, which is a necessity for most applications.

4.2.3.2. INFECTION IN VIVO. One can also deliver virus to the mitotic population in situ using techniques described above for lineage mapping. This may be the method of choice in cases where the mitotic zone is defined and accessible, and culture conditions are unknown. Although the culture conditions must be established eventually, integration and expression of the oncogene prior to explantation of the tissue may make the target cells better able to survive the in vitro conditions. This may be the only way to immortalize cells when the conditions for mitosis in vitro (in the absence of an oncogene) are unknown or are unsuitable. After allowing some time for integration and expression of the vector genome (again, about two cell cycle times), one must then explant the injected tissue and attempt to culture the cells as lines. The drug selection protocol is again applicable in selecting the infected cells.

4.2.4. Characterization of Infected Cells

4.2.4.1. MORPHOLOGICAL AND ANTIGENIC CHARACTERIZATION. As colonies develop under G418 selection, morphology can be observed. We found that it was useful to observe the colonies at several time points, since there were interesting changes that took place as colonies became dense in the center or "matured." In some cases, processes developed, and in other cases, the cells became sparse in the center of the colony and took on a glial morphology, which was not obvious at the edges. It is of course difficult to predict the morphology of progenitor cells if these are not yet defined. However, in some cases, processes, or a particular morphology may prove to be instructive. When morphology is not predictive, or predictable, it is advisable to choose several types of colonies for further characterization.

Colonies can be isolated by cloning cylinders, or initiated by limited dilution techniques, and subsequently analyzed using antibodies. Since this can amount to a great deal of work, antibodies that react with surface molecules can be particularly valuable here. Prior to colony isolation, such antibodies can be used to identify particular cell types on plates containing many colonies. The colonies can be incubated with the antibody, and then a second antibody that is coupled to red blood cells can be used to visualize the positive colonies (Seed and Aruffo, 1987). As this procedure does not harm the cells, it is possible to then isolate positive, viable colonies after this procedure. One can also use a replica plating technique to create copies of the colonies using a filter that supports the growth of cells (Raetz et al., 1982). One copy of the colonies can then be screened by a more noxious method, such as radioactive ligand binding, and then recovery of the viable colony can proceed using the master plate. If such procedures are available, they are worth the time to develop since the effort in isolating, culturing, and screening many colonies is substantial. The importance of a rapid primary screen cannot be overemphasized if a precise, definitive type of line is the object of this procedure.

If no quick screening technique is available, staining colonies individually must be done. We screen on the 8-well Lab-Tek staining slides (Miles). In our experience, many colonies exhibit variable staining results. Within clones, sometimes only 0.001% of the cells will brightly stain with a given monoclonal antibody. In addition, the percentage may vary from staining to staining. In rare instances, the majority of the cells in a given clone stain (and do so reproducibly). These results may suggest that gene expression for different antigens is variable, but independent of commitment or growth status. Alternatively, the cells within a clone vary in their stage of development. Such lines may greatly aid in the understanding of such issues.

4.2.4.2. SCREENING LINES FOR FUNCTION. Clones can be screened for biochemical properties (enzyme activity, excitability, channels) or for a defined function. For example, astrocytes are known to promote the growth of neurons. This assay has been used in a characterization of astrocytic lines, in conjunction with an antigenic characterization (Evrard et al., 1986). Properties of fully differentiated neurons, such as glutamate receptors, have not yet been demonstrated. However, very few lines have been examined for such properties, and attempts to specifically isolate such lines have not been made.

4.2.4.3. ASSAYING LINES IN VIVO. Perhaps the most rigorous assay of a cell's ability to develop and differentiate is an in vivo assay. Since the retroviral marking technique allows one to tag cells with an indelible marker, any clone can be marked via infection with a second retrovirus, such as a lacZ virus. The cells can then be introduced into the animal and later assayed for their behavior (microscopy, isolation followed by in vitro assays, and so on). Alternatively, the viral oncogene may provide an antigenic tag for the cells and they can be identified by using antioncogene antibody (e.g., SV40 T antigen; Frederiksen et al., 1988).

References

Aboud M., Wolfson M., Hassan Y., and Huleihel M. (1982) Rapid purification of extracellular and intracellular Moloney murine leukemia virus. *Arch. of Virology* **71,** 185–195.

Adamson E. D. (1987) Oncogenes in development. *Develop.* **101,** 449–471.

Alt F., Rosenberg N., Lewis S., Thomas E., and Baltimore D. (1981) Organization and reorganization of immunoglobulin genes in A-MuLV-transformed cells: rearrangement of heavy but not light chain genes. *Cell* **27,** 381–390.

Aruffo A. and Seed B. (1987) Molecular cloning of two CD7 (T-cell leukemia antigen)-cDNAs by a COS cell expression system. *EMBO J.* **6,** 3313–3316.

Bartlett P. F., Reid H. H., Bailey K. A., and Bernard O. (1988) Immortalization of mouse neural precursor cells by the c-*myc* oncogene. *Proc. Natl. Acad. Sci. USA* **85,** 3255–3259.

Bender M. A., Palmer T. D., Gelinas R. E., and Miller A. D. (1987) Evidence that the packaging signal of Moloney murine leukemia virus extends into the *gag* region. *J. Virology* **61,** 1639–1646.

Bishop J. M. (1985) Viral oncogenes. *Cell* **42,** 23–38.

Bronner-Fraser M. (1985) Alterations in neural crest migration by a monoclonal antibody that affects cell adhesion. *J. Cell Biol.* **101,** 610–617.

Brown A. M. C. and Scott M. R. D. (1987) Retroviral vectors, in *DNA Cloning: A Practical Approach,* Vol. III, (IRL Press, Oxford and Washington), Ch 9 pp. 189–212.

Brown P. O., Bowerman B., Varmus H. E., and Bishop J. M. (1987) Correct integration of retroviral DNA *in vitro*. *Cell* **49,** 347–356.

Calof A. and Jessell T. (1986) Use of a retroviral vector to transfer and express a foreign gene in murine neurons and neural ectoderm. *Soc. Neurosci. Abst.* **12,** 183.

Cepko C. L. (1988a) Retrovirus vectors and their applications in neurobiology. *Neuron* **1,** 345–353.

Cepko C. L. (1988b) Immortalization of neural cells via oncogene transduction. *Trends Neuro Sci.* **11,** 6–8.

Cepko C. L. (1989) Retrovirus-mediated immortalization of neural cells. *Ann. Rev. Neurosci. Vol. 12* (in press).

Cepko C. L., Roberts B. E., and Mulligan R. E. (1984) Construction and applications of a highly transmissible murine retrovirus shuttle vector. *Cell* **37,** 1053–1062.

Cone R. D. and Mulligan R. C. (1984) High-efficiency gene transfer into mammalian cells: generation of helper-free recombinant retrovirus with broad mammalian host range. *Proc. Natl. Acad. Sci. USA* **81,** 6349–6353.

Cone R. D., Reilly E. B., Eisen H. N., and Mulligan R. C. (1987a) Tissue-specific expression of functionally rearranged λ1 Ig gene through a retrovirus vector. *Science* **236,** 954–957.

Cone R. D., Weber-Benarous A., Baorto D., and Mulligan R. C. (1987b) Regulated expression of a complete human β-globin gene encoded by a transmissible retrovirus vector. *Mol. Cell. Biol.* **7,** 887–897.

Cunningham J., Tseng L., and Scadden D. (1986) Transfer of the gene encoding the murine ecotropic virus receptor into nonpermissive cells, in RNA Tumor Virus Meeting Abstracts 1986 (Eisenman R. and Skalka A. M., eds.), Cold Spring Harbor Laboratory, Cold Spring Harbor, New York, p. 16.

Dannenberg A. M. and Suga M. (1981) in *Methods for Studying Mononuclear Phagocytes* (Adams D. O., Edelson O., and Koren M. S., eds.), Academic Press, New York, pp. 375–396.

Danos O. and Mulligan R. C. (1988) Safe and efficient generation of recombinant retroviruses with amphotropic and ecotropic host ranges. *Proc. Natl. Acad. Sci. USA* **85,** 6460–6464.

Eckhart W. (1969(Complementation and transformation by temperature sensitive mutants of polyoma virus. *Virology* **38,** 120–125.

Embretson J. E. and Temin H. M. (1987) Lack of competition results in efficient packaging of heterologous murine retroviral RNAs and reticuloendotheliosis virus encapsidation-minus RNAs by the reticuloendotheliosis virus helper cell line. *J. Virol.* **61,** 2675–2683.

Emson P. C., Shoham S., Feler C., Price J., and Wilson C. J. (1988) The use of a retroviral vector to identify foetal stratal neurones transplanted into the adult striatum. *Soc. Neurosci. Abstr.* **14,** 1005.

Frederickson K., Jat P. S., Valtz N., Levy D., and McKay R. (1988) Immortalization of precursor cells from the mammalian CNS. *Neuron* **1,** 439–448.

Gardner M. B. (1978) Type C viruses of wild mice: characterization and natural history of amphotropic, ecotropic, and xenotropic MuLV. *Cur. Top. Microbiol. Immunol.* **79,** 215–239.

Geller A. and Breakefield X. O. (1988) A defective HSV-1 vector expresses *Escherichia coli* β-galactosidase in cultured peripheral neurons. *Science* **241,** 1667–1669.

Giotta G. J., Heitzmann J., and Cohn M. (1980) Properties of two temperature-sensitive Rous Sarcoma virus transformed cerebellar cell lines. *Brain Res.* **202,** 445–458.

Glover J. C., Gray G. E., and Sanes J. R. (1987) Patterns of neurogenesis in chick optic tectum studied with a retroviral marker. *Soc. Neurosci. Abstr.* **13,** 183.

Goff S., Traktman P., and Balitmore D. (1981) Isolation and properties of murine leukemia virus mutants: use of a rapid assay for release of virion reverse transcriptase. *J. Virology* **38,** 239–248.

Goring D. R., Rossant J., Clapoff S., Breitman M. L., and Tsui L.-C. (1987) In situ detection of β-galactosidase in lenses of transgenic mice with a γ-crystallin/lacZ gene. *Science* **235,** 456–458.

Gorman C. M., Rigby P. W. J., and Lane D. P. (1985) Negative regulation of viral enhancers in undifferentiated embryonic stem cells. *Cell* **42,** 519–526.

Graham F. L. and van der Erb A. J. (1973) A new technique for the assay of infectivity of human adenovirus 5 DNA. *Virology* **52,** 446–467.

Gray G. E., Glover J. C., Majors J., and Sanes J. R. (1988) Radial arrangement of clonally related cells in the chicken optic tectum: lineage analysis with a recombinant retrovirus. *Proc. Natl. Acad. Sci. USA,* in press.

Hall C. V., Jacob P. E., Ringold G. M., and Lee F. (1983) Expression and regulation of *Escherichia coli lacZ* gene fusions in mammalian cells. *J. Mol. Appl. Genetics* **2,** 101–109.

Hankin M. H. and Lund R. D. (1987) Specific target-directed axonal outgrowth from transplanted embryonic rodent retinae into neonatal rat superior coliculus. *Brain Res.* **408,** 344–348.

Hawley R. G., Covarrubias L., Hawley, T., and Mintz B. (1987) Handicapped retroviral vectors efficiently transduce foreign genes into hematopoietic stem cells. *Proc. Natl. Acad. Sci. USA* **84,** 2406–2410.

Hen R., Dodd J., Cepko C., and Axel R. (1987) Construction of a rat olfactory neuronal cell line. *Soc. Neurosci. Abstr.* **13,** 1410.

Hiromi Y., Kuroiwa A., and Gehring W. (1985) Control elements of the Drosophila segmentation gene *fushi tarazu. Cell* **43,** 603–613.

Hughes S. T., Greenhouse J. J., Petropoulos C. J., and Sutrave P. (1987) Adaptor plasmids simplify the insertion of foreign DNA into helper-independent retroviral vectors. *J. Virology* **61,** 3004–3012.

Jaenisch R. (1980) Retroviruses and embryogenesis: microinjection of Moloney leukemia virus into midgestation mouse embryos. *Cell* **19,** 181–188.

Jaenisch R. and Berns A. (1977) Tumor virus expression during mammalian embryogenesis, in *Concepts in Mammalian Embryogenesis* (Sherman, M., ed.), MIT Press, Cambridge, MA, pp. 267–314.

Jat P. S., Cepko C. L., Mulligan R. D., and Sharp P. A. (1986) Recombinant retroviruses encoding simian virus 40 large T antigen and polyomavirus large and middle T antigens. *Mol. Cell. Biol.* **6,** 1204–1217.

Karlsson S., Papayannopoulou T., Schweiger S. G., Stamatoyannopoulos G., and Neinhuis A. (1987) Retroviral-mediated transfer of genomic globin genes leads to regulated production of RNA and protein. *Proc. Natl. Acad. Sci USA* **84,** 2411–2415.

Korman A. J., Frantz J. D., Strominger J. L., and Mulligan R. C. (1987) Expression of human class II major histocompatibility complex antigens using retrovirus vectors. *Proc. Natl. Acad. Sci. USA* **84,** 2150–2154.

Land H., Chen A. C., Morgenstern J. P., Parada L. F., and Weinberg R. A. (1986) Behavior of *myc* and *ras* oncogenes in transformation of rat embryo fibroblasts. *Mol. Cell. Biol.* **6,** 1917–1925.

Land H., Parada L. F., and Weinberg R. A. (1983) Tumorigenic conversion of primary embryo fibroblasts requires at least two cooperating oncogenes. *Nature (Lond.)* **304,** 596–602.

Largent B. L., Sosnowski R. G., and Reed R. R. (1988) A model for neurogenesis in the mammalian nervous system:generation of an olfactory neuronal cell line. *Soc. Neurosci. Abstr.* **14,** 91.

Lemischka I. R., Raulet D. H., and Mulligan R. C. (1986) Developmental potential and dynamic behavior of hematopoietic stem cells. *Cell* **45,** 917–927.

Levy J. A. (1975) Host range of murine xenotropic virus: replication in avian cells. *Nature (Lond.)* **253,** 140–142.

Lis J. T., Simon J. A., and Stutton C. A. (1983) New heat shock puffs and β-galactosidase activity resulting from transformation of Drosophila with an *hsp70-lacZ* hybrid gene. *Cell* **35,** 403–410.

Lojda Z. (1970) Indigogenic methods for glycosidases. I. An improved method for β -D-glucosidase and its application to localization studies on intestinal and renal enzymes. *Histochemie* **22,** 347–361.

Luskin M. B., Pearlman A. L., and Sanes J. R. (1988) Cell lineage in the cerebral cortex of the mouse studied in vivo and in vitro with a recombinant retrovirus. *Neuron* **1,** 635–647.

Magli M.-C., Dick John E., Huszar D., Bernstein A., and Phillips R. A. (1987) Modulation of gene expression in multiple hematopoietic cell lineages following retroviral vector gene transfer. *Proc. Natl. Acad. Sci. USA* **84,** 789–793.

Mann R., Mulligan R. C., and Baltimore D. (1983) Construction of a retrovirus packaging mutant and its use to produce helper-free defective retrovirus. *Cell* **33,** 153–159.

Markowitz D., Goff S., and Bank A. (1988) A safe packaging line for gene transfer: separating viral genes on two different plasmids. *J. Virology* **62,** 1120–1124.

Miller A. D. and Buttimore C. (1986) Redesign of retrovirus packaging cell lines to avoid recombination leading to helper virus production. *Mol. Cell. Biol.* **6,** 2895–2902.

Miller A. D., Law M-F., and Vermer I. M. (1985) Generation of helper-free amphotropic retroviruses that transduce a dominant-acting, methotrexate-resistant dihydrofolate reductase gene. *Mol. Cell. Biol.* **5**, 431–437.

Mitrani E., Coffin J., Boedtker H., and Doty P. (1987) Rous sarcoma virus is integrated but not expressed in chicken early embryonic cells. *Proc. Natl. Acad. Sci. USA* **84**, 2781–2784.

Mulligan R. C. (1983) Construction of highly transmissible mammalian cloning vehicles derived from murine retroviruses, in *Experimental Manipulation of Gene Expression,* Academic New York, pp. 155–173.

Muneoka K., Wanek N., and Bryant S. V. (1986) Mouse embryos develop normally *exo utero. J. Exp. Zool.* **239**, 289–293.

Nielsen D. A., Chou J., MacKrell A. J., Casadaban M. J., and Steiner D. F. (1983) Expression of a preproinsulin-β-galactosidase gene fusion in mammalian cells. *Proc. Natl. Acad. Sci. USA* **80**, 5198–5202.

Nolan G. P., Fiering S., Nicolas J.-F., and Herzenberg L. A. (1988) Fluorescence-activated cell analysis and sorting of viable mammalian cells based on β-D-galactosidase activity after transduction of E. coli lacZ. *Proc. Natl. Acad. Sci. USA* **85**, 2603–2607.

Norton P. A. and Coffin J. M. (1985) Bacterial β-galactosidase as a marker of Rous sarcoma virus gene expression and replication. *Mol. Cell. Biol.* **5**, 281–290.

Ochman H., Gerber A. S., and Hartl D. L. (1988) Genetic applications of an inverse polymerase chain reaction. *Genetics* **120**, 621–623.

Ow D., Wood K. V., DeLuca M., deWet J. R., Helinski D. R., and Howell S. H. (1986) Transient and stable expression of the firefly luciferase gene in plant cells and transgenic plants. *Science* **234**, 856–859.

Parker B. A. and Stark G. R. (1979) Regulation of Simian virus 40 transcription: sensitive analysis of the RNA species present early in infections by virus or viral DNA. *J. Virol.* **31**, 360–369.

Pearson B., Wolf P. L., and Vazquez J. (1963) A comparative study of a series of new indolyl compounds to localize β-galactosidase in tissues. *Laboratory Investigation* **12**, 1249–1259.

Price J. (1987) Retroviruses and the study of cell lineage. *Development* **101**, 409–419.

Price J. and Thurlow L. (1988) Cell lineage in the rat cerebral cortex: a study using retorviral-mediated gene transfer. *Development,* in press.

Price J., Turner D., and Cepko C. (1987) Lineage analysis in the vertebrate nervous system by retrovirus-mediated gene transfer. *Proc. Natl. Acad. Sci. USA* **84**, 156–160.

Raetz C. R. H., Wermuth M. M., McIntyre T. M., Esko J. D., and Wing D. C. (1982) Somatic cell cloning in polyester stacks. *Proc. Natl. Acad. Sci. USA* **79**, 3223–3227.

Rein A., Schultz A. M., Bader J. P., and Bassin R. H. (1982) Inhibitors of glycosylation reverse retroviral interference. *Virology* **119**, 185–192.
Roberts B. E., Miller J. S., Kimelman D., Cepko C. L., Lemischka I. R., and Mulligan R. C. (1985) Individual adenovirus type 5 early region 1A gene products elicit distinct alterations of cellular morphology and gene expression. *J. Virol.* **56**, 404–413.
Rodriguez J., and Deinhardt F. (1960) Preparation of a semipermanent mounting medium for fluorescent antibody studies. *Virology* **12**, 316–317.
Ryder E., Snyder E., and Cepko C. (1987) Establishment of neural cell lines using retrovirus-vector mediated oncogene transfer. *Soc. Neurosci. Abstr.* **13**, 700.
Ryder E., Snyder E., Deitcher D., and Cepko C. (1988) Establishment and characterization of neural cells via retrovirus-mediated oncogene transduction, in preparation.
Saiki R., Sharf S., Faloona F., Mullis K. B., Horn G. T., Erlich, H. A., and Arnheim N. (1985) Enzymatic amplication of β-globin genomic sequences and restriction site analysis for diagnosis of sickle cell anemia. *Science* **230**, 1350–1354.
Saiki R. K., Gelfand D. H., Stoffel S., Scharf S. J., Higuchi R. G., Horn T. G., Mullis K. B., and Erlich H. A. (1988) Primer-directed enzymatic amplification of DNA with a thermostable DNA polymerase. *Science* **239**, 487–491.
Sanes J. R., Rubenstein J. L. R., and Nicolas J-F. (1986) Use of a recombinant retrovirus to study post-implantation cell lineage in mouse embryos. *EMBO J.* **5**, 3133–3142.
Seed B. and Aruffo A. (1987) Molecular cloning of the CD2 antigen, the T-cell erythrocyte receptor, by a rapid immunoselection procedure. *Proc. Natl. Acad. Sci. USA* **84**, 3365–3369.
Shih C.-C., Stoye J. P., and Coffin J. M. (1988) Highly preferred targets for retrovirus integration. *Cell* **53**, 531–537.
Shimotohno K. and Temin H. M. (1982) Loss of intervening sequences in genomic mouse α-globin DNA inserted in an infectious retrovirus vector. *Nature* **299**, 265–268.
Short M. P., Rosenberg M. B., Ezzedine D., Gage F. H., Friedmann T., and Breakefield E. K. (1988) Autocrine differentiation of rat pheochromocytoma PC12 cells using a retroviral NGF vector. *Soc. Neurosci. Abstr.* **14**, 1115.
Sorge J., Wright D., Erdman V. D., and Cutting A. E. (1984) Amphotropic retrovirus vector system for human cell gene transfer. *Mol. Coll. Biol.* **4**, 1730–1737.

Soriano P. and Jaenisch R. (1986) Retroviruses as probes for mammalian development: allocation of cells to the somatic and germ cell lineages. *Cell* **46**, 19–29.

Soriano P., Cone R. D., Mulligan R. C., and Jaenisch R. (1986) Tissue-specific and ectopic expression of genes introduced into transgenic mice by retroviruses. *Science* **234**, 1409–1413.

Stoker A. and Bissell M. J. (1988) Development of avian sarcoma and leukosis virus-based vector-packaging cell lines. *J. Virology* **62**, 1008–1015.

Tegtmeyer P. (1975) Function of simian virus 40 gene A in transforming infection. *J. Virol.* **15**, 613–618.

Teich N., Weiss R., Martin G., and Lowy D. (1977) Virus infection of murine teratocarcinoma stem cell lines. *Cell* **12**, 973–982.

Turner D. and Cepko C. L. (1987) A common progenitor for neurons and glia persists in rat retina late in development. *Nature (Lond.)* **328**, 131–136.

Turner D. L., Snyder E. Y., and Cepko C. L. (1988) Cell lineage analysis in the embryonic mouse retina by retroviral vector-mediated gene transfer. *Soc. Neurosci. Abstracts,* **14,** 892.

Ursprung H., Sofer W. H., and Burroughs N. (1970) Ontogeny and tissue distribution of alcohol dehydrogenase in *Drosophila melanogaster. Wilhelm Roux' Archiv* **164**, 201–28.

Walsh C. and Cepko C. L. (1988) Clonaly related cortical cells show several migration patterns. *Science* **241**, 1342–1345.

Watanabe S. and Temin H. (1982) Encapsidation sequences for spleen necrosis virus, an avian retrovirus, are between the 5' long terminal repeat and the start of the *gag* gene. *Proc. Natl. Acad. Sci. USA* **79**, 5986–5990.

Watanabe S. and Temin H. (1983) Construction of a helper cell line for avian reticuloendotheliosis virus cloning vectors. *Mol. Cell. Biol.* **3**, 2241–2249.

Weiss R., Teich N., Varmus H., and Coffin J. (1984 and 1985) *RNA Tumor Viruses,* Cold Spring Harbor Laboratory, Cold Spring Harbor, New York.

Werner M., Lieberman P., Madreperla S., and Adler R. (1988) Expression of transfected genes by differentiated, postmitotic neurons and photoreceptors in culture. *Soc. Neurosci. Abstracts* **14**, 623.

Weizacker F. V., Beug H., and Graf T. (1986) Temperature-sensitive mutants of MH2 avian leukemia virus that map in the v-*mil* and v-*myc* oncogene respectively. *EMBO J.* **5**, 1521–1527.

Williams D. A., Lemischka I. R., Nathan D. G., and Mulligan R. C. (1984) Introduction of a new genetic material into pluripotent haematopoietic stem cells of the mouse. *Nature (Lond.)* **310,** 476–480.

Williams D. A., Orkin, S. H., and Mulligan, R. C. (1986) Retrovirus-mediated transfer of human adenosine deaminase gene sequences into cells in culture and into murine hematopoietic cells *in vivo. Proc. Natl. Acad. Sci. USA* **83,** 2566–2570.

Wyke J. A. (1973) The selective isolation of temperature-sensitive mutants of Rouse sarcoma virus. *Virology* **52,** 587–590.

Yu S-F., von Ruden T., Kantoff P. W., Garber C., Seiberg M., Ruther U., Anderson W. F., Wagner E. F., and Gilboa E. (1986) Self-inactivating retroviral vectors designed for transfer of whole genes into mammalian cells. *Proc. Natl. Acad. Sci. USA* **83,** 3194–3198.

Transgenic Mice in Neurobiological Research

Brian Popko, Carol Readhead, Jessica Dausman, and Leroy Hood

1. Introduction

The introduction of cloned genes into cells has greatly increased our knowledge of the factors that regulate the transcription of these genes and of the function of the proteins that they encode. Recently, techniques have been developed that allow the incorporation of cloned genes into the germline of mice (Gordon et al., 1980; Brinster et al., 1981; Costantini and Lacy, 1981). The generation of such transgenic animals has advanced our understanding of the elements involved in regulating tissue specificity and developmental expression of the incorporated genes. Additionally, these studies have yielded insight into the involvement of the transgenes in the physiology of the animal (Palmiter and Brinster, 1986). The application of this technique to genes of neurobiological interest can provide insights into the role specific genes play in the organization and function of the mammalian nervous system. In this article, we provide an overall view of the various aspects involved in generating transgenic animals and some of the potential applications of the technique to genes of neurobiological importance.

2. The Production of Transgenic Mice

The generation of transgenic animals through the microinjection of single-celled embryos with cloned genes involves four main steps: the isolation of fertilized eggs from the oviducts of pregnant animals, the injection of the DNA of interest into one of the pronuclei of the single-celled fertilized embryos, the implantation of the injected embryos into the oviduct of a pseudopregnant female, and the analysis of the generated animals. We will describe each of these steps in limited detail, emphasizing those techniques we most commonly employ. Recently, an excellent laboratory manual has been published (Hogan et al., 1986) that covers all aspects of generating transgenic mice and would be a helpful addition to any laboratory considering setting up for mouse embryo microinjection.

2.1. The Isolation of Fertilized Eggs from the Ampulla of the Oviduct

For most purposes, the female mice that are to act as egg donors are hormonally stimulated to produce a greater yield of embryos. Unstimulated females of most strains produce about 8–12 eggs per ovulation, whereas hormonally stimulated females will produce 15–40 eggs (Hogan et al., 1986). About 48 h before the matings are to be set up, the animals are injected with 5 international units (IU) intraperitoneally of pregnant mare's serum (PMS), which stimulates follicular growth. Forty-eight hours later, the animals are again injected intraperitoneally with 5 IU of human chorionic gonadotropin (hCG), which potentiates the resumption of oocyte meiosis and ovulation. Upon hormonal stimulation, some strains do not produce a significantly higher yield of embryos. Nevertheless, even in strains that do not respond well, a higher percentage of hormonally stimulated females plug than females in natural matings. Although hormone-stimulated matings will produce a larger number of abnormal eggs than normal matings, the number of injectable eggs produced per female is generally greatly increased.

In order to obtain consistent results, it is important to maintain the mice in a well-controlled environment. One of the most important environmental considerations is the animal's light–dark cycle. Our mice are maintained on a daily cycle of 12 h of light and 12 h of dark. When the dark period is between 4 PM and 4 AM, fertilization occurs around midnight and the embryos are isolated the following morning. At this stage, the eggs are surrounded by cumulus cells and are localized within the swollen ampulla of the oviduct (Fig. 1). The oviduct is dissected away from the ovary and uterus and placed in M2 culture medium (Hogan et al., 1986). The eggs are recovered under a dissecting microscope at 10–20× magnification

Transgenic Mice 223

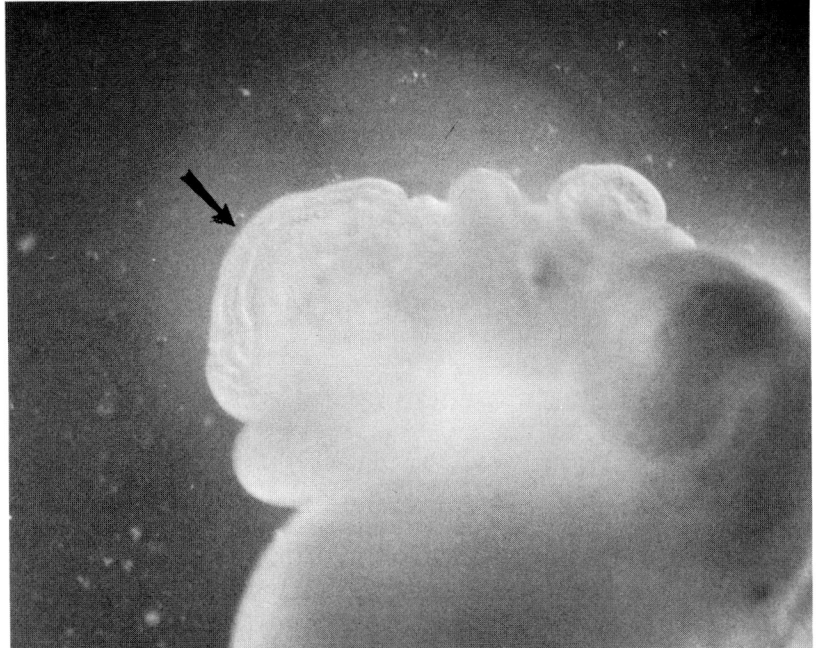

Fig. 1. The oviduct of a pregnant female mouse. This photograph shows the oviduct approximately 10 h after fertilization. The arrow is pointing to the swollen ampulla, which contains the fertilized eggs.

by tearing the oviduct near the swollen ampulla. The eggs are pipetted into a solution of 0.1% hyluronidase to remove the cumulus cells. After the cumulus cells have been completely removed, the eggs are washed in M2 medium and then placed in droplets of M16 medium (Hogan et al., 1986) under paraffin oil and incubated at 37°C in 5% CO_2 until they are injected.

2.2. The Injection of DNA into the Pronuclei of Fertilized Eggs

The DNA to be microinjected needs to be purified extensively to remove any contaminating substances that might either be harmful to the egg or clog the injection needle. Because the most desirable DNA form to be injected is a linear molecule (Brinster et al., 1985; Hammer et al., 1987) that has been freed of prokaryotic vector sequences (Chada et al., 1985; Townes et al., 1985), the purification process must remove unwanted protein (e.g., restriction enzymes) and DNA (vector sequences). The most common purification process that is carried out in our laboratory involves the use of sucrose gradients. The DNA is digested with the appropriate restriction enzyme to remove as much of the vector sequences as possible and then extracted with phenol to remove the enzymes. Following an ethanol precipitation, the DNA is layered on a linear sucrose gradient (usually 10–30%, but this can be modified depending on the sizes of the DNA fragments to be resolved). After centrifugation, fractions are collected and analyzed. Those that contain the appropriate DNA fragment are dialyzed extensively. The final purification step is a long spin (at least 30 min) in a microcentrifuge (= 16,000 × g) to remove any particulate matter that might clog the injection needle. The concentration of the DNA is then adjusted such that approximately 200 molecules are injected in one picoliter (1 ng/µL for a 5-kb DNA molecule).

Microinjection setups vary from lab to lab, but the basic components are shared by most. The central item to the microinjection process is the injection microscope. Most laboratories use an inverted microscope with either Nomarski differential interference contrast optics or Hoffman modulation contrast optics. The injection of the egg is carried out at about 200× magnification. The microscope should be situated in a location free from vibration or on a vibration-free table to ensure that the egg to be injected is a stable target and that the injection needle remains steady.

The holding and injection pipets are each controlled by micromanipulators that are situated on either side of the microscope. The egg is held in place by slight suction on the holding pipet (Fig. 2) that is attached to a micrometer, via plastic tubing that is filled with a biologically inert liquid. On the opposite side of the microscope, the injection needle is connected with air-filled tubing to a large syringe that controls the DNA injection. The injections are carried out by first bringing one of the pronuclei of the single-celled egg into sharp focus using the adjustment knobs of the microscope. Next the injection needle is brought into the same focal plane as the pronucleus by using the micromanipulator of the needle to bring it into focus. The joystick of the needle's micromanipulator is used to place the tip of the needle within the pronucleus, and pressure on the syringe is used to inject about one picoliter of the DNA solution. After the eggs are injected, they are returned to M16 medium under oil and incubated at 37°C with 5%

Transgenic Mice

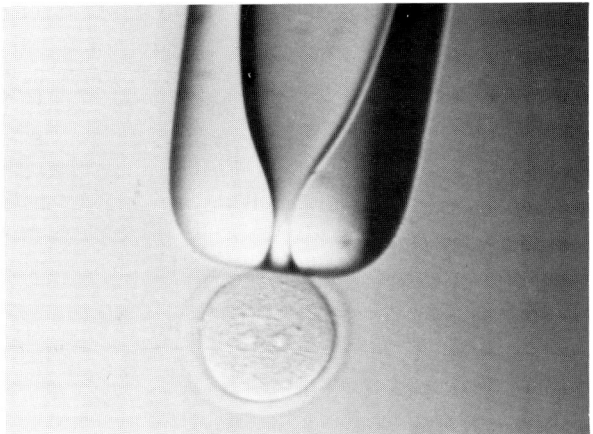

Fig. 2. The injection of DNA into the pronuclei of a single-celled mouse embryo. The embryo is being held in place by the holding pipet. The male and female propnuclei can be seen in the center of the cell.

CO_2 overnight. The embryos that have divided to the two-cell stage by the next day are transferred to the oviduct of a pseudopregnant female. Alternatively, the eggs may be transferred immediately following injection to the pseudopregnant female. In our experience, however, it is better to allow the embryos to incubate overnight to ensure that only viable two-celled embryos are transferred. Approximately 50% of the embryos we have injected progressed to the two-cell stage (Table 1).

2.3. The Transfer of Injected Embryos into the Oviduct of Pseudopregnant Females

Those embryos that survive injection are transferred to the ampulla of the oviduct of a pseudopregnant female through the infundibulum. Pseudopregnant females are generated by mating vasectomized (i.e., sterile) males to the females the night before the transfers are to take place. Plugged females are anesthetized by an intraperitoneal injection of Avertin (Hogan et al., 1986), which is an anesthetic of medium duration (about 1 h) and a light analgesic. A small dorso-lateral incision is made through the skin and body wall across the left side of the lower back of the animal. The ovary is attached to a fat pad that can easily be grasped with forceps. The fat

Table 1
Compilation of Data from the Caltech Microinjection Facility[1]

Manipulation	Number/total analyzed	Percentage
Eggs developing to two-cell embryos after injection[2]	7049/14000	50%
Mice born from transferred two-cell embryos[3]	1088/7049	15%
Transgenic mice among offspring[4]	100/1088	9%
Transgenic mice expressing integrated gene[5]	24/53 (tested)	45%

[1]This data represents the work of ten researchers injecting 24 constructs over a period of 2 y. Most of the researchers have never used this technique before, and therefore, the data from their first 3 m have not been included.

[2]The survival of injected eggs to the two-cell stage in culture is dependent on the skill of the injector and the strain of mice used as egg donors. Some eggs such as those from shiverer mice lyse very easily. The eggs used in these experiments came from B6D2F1/J, Swiss, C3H/HeJ, C57BL/6J, BrSJLF1/J, B6SJLF1/J, and shiverer donors.

[3]In all cases, B6D2F1/J pseudopregnant females were used as foster mothers.

[4]Successful integration of the injected DNA into the genome depends on a number of factors. In our experience, the method of purification of the DNA and linearization of the molecule seem to be among the most important.

[5]Expression of the foreign gene depends on a number of factors. These factors include: site of integration, presence of vector sequences, sufficient flanking sequence, and compatible promoter and enhancer elements (Palmiter and Brinster, 1986).

pad is pulled across the back of the mouse, exposing the oviduct. Under a dissecting microscope with about 20× magnification, the infundibulum of the oviduct can be seen through the ovarian bursa. This transparent membrane must be torn in order for the transfer pipet to enter the infundibulum. The embryos are taken up into the transfer pipet in a minimum volume and inserted into the infundibulum (Fig. 3). A convenient way of assuring that there has been a successful transfer is to continue blowing until an air bubble

Transgenic Mice

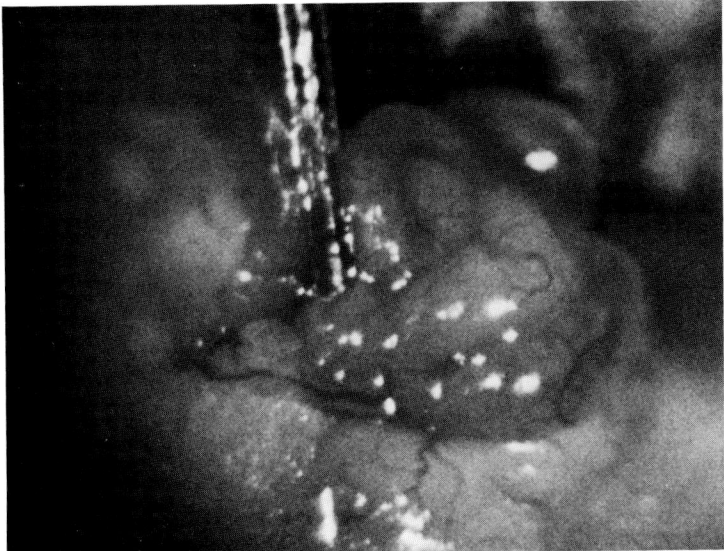

Fig. 3. The transfer of injected embryos into the oviduct of a pseudopregnant mouse. This photograph shows the oviduct as the eggs are being transferred into the infundibulum through the transfer pipet.

appears in the ampulla. After the transfer, the uterus, oviduct, and ovary are carefully placed back inside the animal and the incision closed. We generally transfer about 15–20 eggs to one oviduct per female. The eggs will migrate into the opposite uterine horn and thus implant on both sides (Dausman, unpublished observations). In our experience, slightly less than 15% of all injected eggs will be successfully transferred and develop into pups (Table 1).

2.4. The Analysis of Pups

If the embryo transfer is successful, pups will be born about 20 d later. The animals are examined for the presence of the transgene at 2–3 w of age. DNA is isolated from a small portion of the tail using one of several procedures (Hogan et al., 1986). We generally digest the minced tail sample in proteinase K overnight, followed by several phenol-chloroform extractions and an ethanol precipitation. The purified DNA is usually clean enough for dot blot or restriction-enzyme analysis. Animals identified as having the transgene are mated, and their progeny examined for the presence and expression of the transgene. Approximately 10% of the animals we have generated have been transgenic (Table 1). Founder animals are usually not used for expression studies because of the possibility that the DNA integrated after the embryo divided to the two-cell stage and the animal's cells are heterogenous with respect to the presence of the transgene (i.e., mosaic). When an interesting line of transgenic animals is generated, animals homozygous for the transgene are usually established to eliminate the need of identifying transgenic progeny resulting from the mating of heterozygotes.

2.5. The Management of Data by Computer

When transgenic mice are generated with several constructs and when several lines of trangenic animals are maintained, the management of data and of the mouse colony become an ardous task. Consequently, a computer program (Mousetree) has been developed to assist in this process. Mousetree is a simple graphical database program written specifically for transgenic experiments by Douglas Whiting of Radix Instruments. This program runs on a IBM PC or compatible computer with 256K memory, DOS 2.0 or greater, and a single disk drive, although a hard disk is recommended for storing the database. Transgenic mice and their offspring are linked graphically by a "tree," parts of which appear on the screen. By moving the cursor, one is able to highlight the mouse of interest or move around the tree to view the rest of the family. When a mouse is highlighted, an information card appears on the right-hand side of the screen, and data relevant to that mouse may be written and stored on this card and updated whenever necessary. The format of the card is not fixed, and thus, the researchers are able to store information that suits their needs. Additionally, it is possible to store information about the injected gene constructs at the "root" of the tree. Mousetree has extended commands that enable the user to select subsets of data from the information cards. These data may either be summed or the relevant cards may be printed. This program cannot be used for complex statistical analysis; nevertheless, we have found it is useful for managing transgenic mouse experiments and the mouse colony.

3. Other Approaches of Introducing Foreign DNA into the Germline of Mice

3.1. The Generation of Transgenic Mice Through the Use of Pluripotent Stem Cells

An alternative route to the generation of transgenic mice is through the use of pluripotent embryonic stem (ES) cells (Hooper et al., 1987; Kuehn et al., 1987). These are cell lines that grow in culture and, when injected into normal embryos, have the potential to contribute to all tissues of the developed animal (including germ cells) (Martin, 1980). ES cells are injected into the inner cell mass at the blastula stage, and the embryo is implanted into the uterus of a pseudopregnant animal. The chimeric mice that develop vary in the proportion of cells that are contributed by the ES cells. A stable line of animals with the ES cells genotype is produced when the ES cells contribute to the germline.

A gene of interest can be introduced into the ES cells in culture, and when these cells are injected into embryos, transgenic animals are generated. Although this technique is more cumbersome than the microinjection of DNA directly into fertilized eggs, there are several situations where ES cells are the route of choice for the generation of transgenic mice. For example, ES cells were recently utilized to generate mice defective in the gene encoding hypoxanthine-guanine phosphoribosyl transferase (HPRT) (Kuehn et al., 1987). ES cells were infected with retroviruses that acted as random insertional mutagens (Robertson et al., 1986) to inactivate the HPRT gene. ES cells with the HPRT$^-$ phenotype were selected in culture, and when these cells were injected into mouse blastocysts, HPRT$^-$ mice were produced (Kuehn et al., 1987). Hooper et al. (1987) utilizing HPRT$^-$ ES cells that arose through spontaneous mutational events have also generated HPRT$^-$ mice. A second example of the utility of ES cells is when a gene is being examined whose product would be lethal if expressed in all cells. Because the original animal produced when ES cells are injected into blastocysts is a chimera, the effect of a lethal gene can be examined on a wild-type background.

3.2. The Use of Retroviruses to Introduce Foreign Genetic Material into the Germline of Mice

Retroviruses have also been utilized to introduce foreign genetic material into the germline of mice (Jaenisch, 1976). Early embryos are exposed to infectious virus and then transferred to the uterus of a pseudopregnant mouse. Occasinally, mice are generated that have the viral DNA integrated into the genome of their germinal cells. This method of generating transgenic mice is less cumbersome than the microinjection of eggs or the manipulation of embryonic stem cells. Nevertheless, the construction of the viral vector containing the gene of interest requires considerable effort, and the size of insert that can be incorporated into retroviral vectors is limited by the size of the RNA molecule that can be packaged into the virion. The technique has, however, recently been utilized to generate mice expressing foreign genes in a tissue-specific manner (Soriano et al., 1986).

4. The Application of the Technique of Generating Transgenic Mice to Questions of Neurobiological Interest

4.1. The Elucidation of Factors Involved in Regulating Gene Expression

Since 1980, when the first transgenic animals were generated by microinjection, a large number of transgenic studies have been carried out that have examined the cis-acting elements regulating tissue and temporal specificity of gene expression (Palmiter and Brinster, 1986). The results of these studies have demonstrated that, in most cases, these controlling elements are located within a few kb either upstream or downstream of the coding region of the gene. Similar studies with genes expressed in specific cell types of the nervous system should be equally informative. Because there are few cell lines that share many of the characteristics of differentiated neural or glial cells, it is impossible to assay for cis-regulatory elements in vitro for most genes of neurobiological interest. Therefore, the best way to define these elements is to utilize transgenic technologies. Although it appears unlikely that neural genes will prove to employ radically different regulatory mechanisms, these studies will contribute valuable information for generating constructs that will express reporter molecules to specific neural cell types *(see below)*.

4.2. The Ablation of Specific Cell Lineages

Recently, Palmiter and his coworkers demonstrated a new and powerful application of transgenic technology (Palmiter et al., 1987). Using the transcriptional regulatory region of the mouse elastase gene, they were able to promote the transcription of the toxin produced by the Diptheria bacterium solely in the secretory cells of the pancreas of mice transgenic for the gene construct. The production of this toxin selectively ablated those cells of the pancreas that normally synthesize elastase, but did not harm surrounding cells or other cells that do not normally synthesize elastase. The application of this technique to cells of the nervous system could produce interesting and informative lines of mice. The central nervous system is composed of an extreme multitude of cell types, and this has contributed to the difficulty of associating given functions with the cells responsible. Consequently, as more cell-specific genes are isolated, the application of the cell ablation technique should provide a powerful tool in dissecting the functions of the nervous system.

4.3. The Transformation of Specific Cell Types In Vivo

The examination of cell lines that share many of the properties of the cells from which they were derived has provided insight into the function of the normal cells. Additionally, the analysis of a series of transformed cells that have originated from multiple points along a cell lineage has proven useful in the study of the biochemical processes associated with the differentiation process (Alt et al., 1987). Consequently, the search for cell lines that share properties of the progenitor cells for terminally differentiated glial and neural cells has been extensive (Nelson, 1977; Pfeiffer et al., 1977). A common approach to this search has been to mutagenize an animal with the hope that neoplastic cells of an interesting origin will develop.

With the introduction of transgenic technologies, it is now possible to transform specific cell types in vivo. This is accomplished by utilizing the transcriptional regulatory region of a gene that is expressed specifically in the cell of interest to promote the transcription of an oncogene. The abnormal expression of the oncogene leads to the transformation of the cell. Hanahan (1985) has transformed the insulin-secreting β cell of the pancreas by utilizing the transcriptional regulatory region of the insulin gene to promote the transcription of the simian virus 40 oncogene. Additionally, others (Stewart et al., 1984; Adams et al., 1985; Mahon et al., 1987) have used this approach to generate transformed cells of defined origin. Similar studies using the transcriptional regulatory regions of genes expressed in specific neural and glial cells should result in transformed cells with interesting phenotypes.

4.4. The Generation of Mouse Mutants of Defined Origin

Perhaps the most exciting potential application of transgenic mice is in the inactivation of genes. The examination of mutants, from prokaryote to higher mammals, has provided us with vast insights into the function of the proteins encoded normally by the genes affected by the mutation. Additionally, the inactivation of genes known to be responsible for genetic deseases in humans could result in animal models of these diseases. Consequently, a considerable amount of effort has been expended trying to devise ways to inactivate genes in higher eukaryote cells to generate mutants of defined origin. The most common approach to the inactivation of genes has been to transfect cells with cloned genes in ways to promote homologous recombination between the cloned genes and the endogenous genes. Although homologous recombination occurs at frequencies considerably higher than random integration (Smithies et al., 1985; Doetschman et al., 1987; Thomas and Capecchi, 1987), the frequency is still not high enough for this to be a feasible approach to gene inactivation in transgenic mice. Nevertheless, intense efforts are being made to find ways of promoting homologous recombination to frequencies suitable for transgenic studies.

A second technique of gene inactivation that has recently gained considerable attention utilizes antisense RNA to inhibit translation of the transcripts of the gene of interest. The antisense RNA presumably hybridizes to the sense transcripts and thereby inhibits their translation. When cells are transfected with constructs that promote the transcription of the antisense RNA of the thymidine kinase (tk) gene, tk enzyme activity is suppressed (Izant and Weintraub, 1985; Kim and Wold, 1985). It is possible that the expression of antisense transcripts in transgenic mice will result in the generation of mice deficient in the protein of interest.

Herskowitz (1987) has recently proposed an alternative strategy to the functional inactivation of genes. He suggests that the overproduction of a mutant form of the protein of interest will result in the abnormal protein out-competing the endogenous protein and the production of a mutant phenotype. Although this strategy has not been attempted experimentally, it appears to occur naturally in humans (Prockop, 1984). Lethal osteogenesis imperfecta is a dominant mutation that is the result of the incorporation of a mutant subunit of collagen into the polyprotein chain. Consequently, this approach to the generation of mouse mutants may be feasible, especially when examining structural proteins that are involved in protein–protein interactions.

Another mechanism by which genes may be inactivated is through random insertional mutagenesis. Occasionally, when foreign DNA integrates into the fertilized egg's genome, a gene of the host cell is disrupted. It has been estimated that approximately 7% of all transgene integrations result in insertional mutagenesis (Palmiter and Brinster, 1986). Usually, these mutations are recessive, and consequently, no phenotypic effect is observed until the animals are made homozygous for the transgene. Although the foreign DNA integrates randomly and little information is immediately available as to the molecular defect associated with the observed phenotype, the transgene acts as a molecular tag that can be used to isolate quickly the region of the genome that has been disrupted by the transgene. It is likely that genes of neurological interest will occasionally be randomly mutated by this mechanism.

In addition to these applications, individual genes will provide alternative, specialized applications of the transgenic technique. For example, some proteins may be specially suited to study structure–function relationships in vivo. Alternatively, certain proteins may provide insight when expressed in an incorrect tissue or temporal-specific manner. Table 2 describes a number of transgenic studies that have already provided interesting results with applications to the nervous systems. The continued application of the transgenic technique to genes of neurobiological interest should provide informative results unattainable through conventional means.

Table 2
Expression of Introduced Genes in the Nervous System of Transgenic Mice

Structural gene sequences	Regulatory sequences	Results
SV40 early region	Metallothionein(MT)	Choroid plexus tumors (Brinster et al., 1984)
SV40 T antigen	SV40 enhancer	Choroid plexus tumors (Palmiter et al., 1985)
SV40 T antigen	mT/human growth hormone fusion	Peripheral neuropathies (gliomas) (Messing et al., 1985)
JC virus early region	JC virus early region	Central nervous system dysmyelination (Small et al., 1986a) and adrenal neuroblastomas (Small et al., 1986b)
Myelin basic protein (MBP)	MBP	Correction of the dysmyelinating phenotype of shiverer (Readhead et al., 1987) and myelin-deficient (Popko et al., 1987) mice
Gonadotrophin-releasing hormone (GnRH)	GnRH	Restoration of fertility in the hypogonadal (hpg) mouse (Mason et al., 1986)

Acknowledgments

We thank Richard Barth, Jane Johnson, Elsebet Lund, Murray Robinson, Frank Sangiorgi, Tow Wilkie, and Carol Wuenschall for contributing to the data presented in Table 1. We also thank Ulf Landegren and Vipin Kumar for critically reviewing the manuscript, and Cathy Elkins for preparing the manuscript. The Caltech Microinjection Facility is supported by grants from the NIH.

References

Adams J. M., Harris A. W., Pinkert C. A., Corcoran L. M., Alexander W. S., Cory S., Palmiter R. D., and Brinster R. L. (1985) The c-Myc oncogene driven by immunoglobin enhancers induces lymphoid malignancy in transgenic mice. *Nature (Lond.)* **318**, 533–538.

Alt F. W., Blackwell T. K., and Yancopoulos G. D. (1987) Development of primary antibody repertoire. *Science* **238**, 1079–1081.

Brinster R. L., Chen H. Y., Trumbauer M. E., Senear A. W., and Warren R., and Palmiter R. D. (1981) Somatic expression of herpes thymidine kinase in mice following injection of a fusion gene into eggs. *Cell* **27**, 223–231.

Brinster R. L., Chen H. Y., Messing A., van Dyke T., Levine A. J., and Palmiter R. D. (1984) Transgenic mice harboring SV40 T-antigen genes develop characteristic brain tumors. *Cell* **37**, 367–379.

Brinster R. L., Chen H. Y., Trumbauer M. E., Yagle M. K., and Palmiter R. D. (1985) Factors affecting the efficiency of introducing foreign DNA into mice by microinjecting eggs. *Proc. Natl. Acad. Sci. USA* **82**, 4438–4442.

Chada K., Magram J., Raphael K., Radice G., Lacy E., and Costantini F. (1985) Specific expresion of a foreign β-globin in erythroid cells of transgenic mice. *Nature (Lond.)* **314**, 377–380.

Costantini F. and Lacy E. (1981) Indroduction of a rabbit β-globin gene into the mouse germ-line. *Nature (Lond.)* **294**, 92–94.

Doetschman T., Gregg R. G., Maeda N., Hooper M. L., Melton D. W., Thompson S., and Smithies O. (1987) Targeted correction of a mutant HPRT gene in mouse embryonic stem cells. *Nature (Lond.)* **330**, 576–578.

Gordon J. W., Scangos G. A., Plotkin D. J., Barbosa J. A., and Ruddle F. H. (1980) Genetic transformation of mouse embryos by micoinjection of purified DNA. *Proc. Natl. Acad. Sci. USA* **77**, 7380–7384.

Hammer R. E., Krumlauf R., Camper S. A., Brinster R. L., and Tilghman S. M. (1987) Diversity of alpha-fetoprotein gene expression in mice is generated by a combination of separate enhancer elements. *Science* **235**, 53–58.

Hanahan D. (1985) Heritable formation of pancreatic β-cell tumors in transgenic mice expressing recombinant insulin/simian virus 40 oncogenes. *Nature (Lond.)* **315**, 115–122.

Herskowitz I. (1987) Functional inactivation of genes by dominant negative mutations. *Nature (Lond.)* **329**, 219–222.

Hogan B., Costantini F., and Lacy E. (1986) *Manipulating the Mouse Embryo* (Cold Spring Harbor Laboratory, New York, N.Y.).

Hooper M., Hardy K., Handyside A., Hunter S., and Monk M. (1987) HPRT-deficient (Lesch-Nyhan) mouse embryos derived from germline colonization by cultured cells. *Nature (Lond.)* **326**, 292–295.

Izant J. G. and Weintraub H. (1985) Constitutive and conditional suppression of exogenous and endogenous genes by anti-sense RNA. *Science* **229**, 345–352.

Jaenisch R. (1976) Germ line integratin and Mendelian transmission of the exogenous Moloney leukemia virus. *Proc. Natl. Acad. Sci. USA* **73**, 1260–1264.

Kim S. and Wold B. J. (1985) Stable reduction of thymidine kinase activity in cells expressing high levels of anti-sense RNA. *Cell* **42**, 129–138.

Kuehn M. R., Bradley A., Robertson E. J., and Evans M. J. (1987). A potential animal model for Lesch-Nyhan syndrome through introduction of HPRT⁻ mutations ito mice. *Nature (Lond.)* **326**, 295–298.

Mahon K. A., Chepelinsky A. B., Khillan J. S., Overbeck P. A., Piatigorsky J., and Westphal H. (1987) Oncogenesis of the lens in trangenic mice. *Science* **235**, 1622–1628.

Martin G. R. (1980) Teratocarcinomas and mammalian embryogenesis. *Science* **209**, 768–775.

Mason A. J., Pitts S. L., Nikolics K., Szonyi E., Wilcox J. N., Seeburg P. H., and Stewart T. A. (1986) The hypogonadal mouse: Reproductive functions restored by gene therapy. *Science* **234**, 1372–1378.

Messing A., Chen H. Y., Palmiter R. D., and Brinster R. L. (1985) Peripheral neuropathies, hepatocellular carcinomas and islet cell adenomas in transgenic mice. *Nature (Lond.)* **316**, 461–463.

Nelson P. L. (1977) *Neuron Cell Lines, in Cell, Tissue and Organ Cultures in Neurobiology* (Federoff S. and Hertz L, eds.), Academic Press, New York, pp. 347–365.

Palmiter R. D. and Brinster R. L. (1986) Germ-line transformation of mice. *Ann. Rev. Genet.* **20**, 465–499.

Palmiter R. D., Behringer R. R., Quaife C. J., Maxwell F., Maxwell I. H., and Brinster R. L. (1987) Cell lineage ablation in transgenic mice by cell-specific expression of a toxin gene. *Cell* **50**, 435–443.

Palmiter R. D., Chen H. Y., Messing A., and Brinster R. L. (1985) SV40 enhancer and large T-antigen are instrumental in development of choroid plexus tumors in transgenic mice. *Nature (Lond.)* **316**, 457–460.

Pfeiffer S. E., Betschart B., Cook J., Mancini P., and Morris R. (1977) *Glial Cell Lines, in Cell, Tissue, and Organ Cultures in Neurobiology* (Federoff S. and Hertz L., eds.) Academic Press, New York, pp. 287–346.

Popko B., Puckett C., Lai E., Shine H. D., Readhead C., Takahashi N., Hunt S. W., Sidman R. L., and Hood L. (1987) Myelin-deficient mice:

Expression of myelin basic protein and generation of mice with varying levels of myelin. *Cell* **48,** 713–721.

Prockup D. J. (1984) Osteogenesis imperfecta: phenotypic heterogeneity, protein suicide, short and long collagen. *Am. J. Hum. Genet.* **36,** 499–505.

Readhead C., Popko B., Takahashi M., Shine H. D., Saavedra R. A., Sidman R. L., and Hood L. (1987) Expression of myelin basic protein gene in transgenic shiverer mice: Corrections of the dysmyelinating phenotype. *Cell* **48,** 703–712.

Robertson E., Bradley A., Kuehn M., and Evans M. (1986) Germ-line transmission of genes introduced into cultured pluripotential cells by retroviral vector. *Nature (Lond.)* **323,** 445–448.

Small J., Scangos G. A., Cork L., Jay G., and Khoury G. (1986a) The early region of human papovavirus JC induces dysmyelination in transgenic mice. *Cell* **46,** 13–18.

Small J. A., Khoury G., Jay G., Howley P. M., and Scangos G. A. (1986b) Early regions of JC virus and BK virus induce distinct and tissue-specific tumors in transgenic mice. *Proc. Natl. Acad. Sci. USA* **83,** 8288–8292.

Smithies O., Gregg R. G., Boggs S. S., Koralewski M. A., and Kucherlapati R. S. (1985) Insertion of DNA sequences into the human chromosomal β-globin locus by homologous recombination. *Nature (Lond.)* **317,** 230–234.

Soriano P., Cone R. D., Mulligan R. C., and Jaenisch R. (1986) Tissue-specific and ectopic expression of genes introduced into transgenic mice by retroviruses. *Science* **234,** 1409–1413.

Stewart T. A., Pattengale K., and Leder P. (1984) Spontaneous mammary adenocarcinomas in transgenic mice that carry and express MTV/myc fusion genes. *Cell* **38,** 627–637.

Thomas K. R. and Capecchi M. R. (1987) Site-directed mutagenesis by gene targeting in mouse embryo derived stem cells. *Cell* **51,** 503–512.

Townes T. M., Chen H. Y., Lingrel J. B., Palmiter R. D., and Brinster R. L. (1985) Expression of human β-globin genes in transgenic mice: effects of a flanking metallothionein-human growth hormone fusion gene. *Mol. Cell. Biol.* **5,** 1977–1983.

In Situ Hybridization

Michael C. Wilson and Gerald A. Higgins

1. Introduction

The application of conventional hybridization methods to detect gene sequences and their messenger RNAs in cytochemical and histological preparations was introduced 20 years ago. With the aid of several biological systems to provide model systems and abundant gene products, these early studies established *in situ* hybridization as a sensitive technique to resolve and quantitate nucleic-acid sequences at the cellular level. For example, the first demonstrations of *in situ* hybridization to genomic DNA, pioneered by Gall and Pardue (1969), John et al. (1969), and Buongiorno-Nardelli and Amaldi (1970), used in vivo labeled ribosomal RNA as a probe to visualize the several hundred reiterated copies of ribosomal genes clustered within the nucleolus of a variety of cell lines and tissues, including brain. In the early exploitation of viral systems, *in situ* hybridization was made feasible by the availability of viral genomes to provide probes and was used to demonstrate successfully viral sequences in individual tumor cells (Orth et al., 1970; zur Hausen and Schulte-Holthausen, 1972; Dunn et al., 1973). Similarly, the ability to synthesize complementary DNA (cDNA) from an enriched fraction of 9S globin mRNA with viral reverse transcriptase allowed for the first detection of a cellular mRNA (Harrison et al., 1973). With availability of recombinant cDNA and synthetic oligonucleotide probes, *in situ* hybridization is no longer limited to viral and other specialized systems. Although modifications to improve sensitivity and resolution are still being introduced, the use of *in situ* hybridization already has been extended to many fields of biological study, and particularly to the neurosciences, where it provides an important tool to examine gene expression at the cellular level.

2. Applications of *In Situ* Hybridization to Problems in Neurobiology

In situ hybridization of mRNA has now become part of the standard repertoire of neurobiological techniques (Uh1, 1986; Valentino et al., 1987). Because this method allows the direct visualization of the distribution of steady-state levels of specific RNA transcripts, it reveals the extraordinary cellular complexity of the mammalian brain in the context of gene expression. As an approach to the neurosciences, it can be applied to problems which were once considered intractable to more conventional analytic methods. For example, it may be used to study changes in gene expression within individual neurons following a variety of stimuli. In particular, it may be used at great advantage in those situations, such as central nervous system (CNS) development and quantitative analysis, where technical or biological concerns limit the utility of more traditional immunocytochemical approaches. *In situ* hybridization can also complement immunocytochemical studies that describe the cell phenotype defined by translational and post-translational events. Although primarily determined at the level of transcription, the abundance of mRNA can be influenced by the efficiency, and differential processing of primary nuclear mRNA precursors as well as by the cytoplasmic stability of the mRNA (Darnell and Wilson, 1982). *In situ* hybridization, therefore, provides the means to begin to explore how gene expression is regulated during the development of the CNS and as a consequence of normal as well as pathological brain function.

Some of the first applications of *in situ* hybridization to the mammalian nervous system were initiated by neuroendocrinologists for investigations of hormone gene expression and regulation in the pituitary gland. The hypophysis provided a model system in which to study morphologically circumscribed cell types with well-characterized, abundant mRNA sequences such as growth hormone and proopiomelanocortin (POMC), and to examine the technical parameters of the method in a tissue reminiscent of the CNS (Hudson et al., 1981; Gee and Roberts, 1983). Indeed, the earliest report of *in situ* hybridization of a mRNA species in the CNS was the report of Gee et al. (1983), who identified POMC mRNA within ACTH-containing neurons in the periarcuate region of the rat hypothalamus with a ^{3}H-cDNA probe. With synthetic oligonucleotides, cDNA, and cRNA probes, numerous studies have extended the use of *in situ* hybridization to localize mRNAs associated with other neurotransmitter systems to specific cells within the CNS, including mRNAs encoding neuropeptides and neurotransmitter biosynthetic enzymes (Fig. 1) (Uhl, 1986; Valentino et al., 1987), as well as neurotransmitter receptors (Goldman et al., 1986).

One limitation in using *in situ* mRNA hybridization to analyze gene expression is that it measures steady-state levels of cytoplasmic mRNA that may be slow to respond to rapid changes in gene transcription. A more direct approach to detect transcriptional regulation would be to identify directly the primary gene transcripts, i.e., heterologous nuclear RNA products of RNA polymerase II. Fremeau et al. (1986) used an intron-specific probe to identify POMC heterologous nuclear RNA (hnRNA) transcripts within individual-cell nuclei of the rat pituitary. They were able to detect glucocorticoid-induced inhibition of short-lived POMC hnRNA transcripts within corticotrophs, prior to subsequent changes in the cytoplasmic levels of the corresponding mRNA product. With the use of this model system, further optimization of the sensitivity of *in situ* hybridization may allow application of the technique for the study of rapid changes in gene expression within central neurons in response to chemical or electrical signals.

One application of *in situ* hybridization is the study of gene expression in the developing CNS. For example, it may be possible to identify cells or regions expressing a particular mRNA prior to the elaboration of the mature translation product. In developing rat brain, Fink et al. (1986) were able to detect somatostatin mRNA within the forebrain and ventral horn of the spinal cord at early embryonic ages (E12), days before the appearance of somatostatin peptide within these regions (Nobou et al., 1985; Fink et al., 1986). In addition, it may be possible to study the very early expression of genes involved in early determinative events in an analogous fashion to studies in well-characterized developmental systems such as pattern formation in Drosphila (Akam, 1983; Hafen et al., 1983). Recent studies suggest that specific homeobox genes may be expressed in circumscribed regions of the developing spinal cord similar to the segmental expression of homologous genes in Drosophila (Utset et al., 1987). The technique can also be used as part of a more general strategy to identify those genes that are expressed at high abundance during specific developmental stages of the CNS. For example, Miller et al. (1987) have used *in situ* hybridiza-

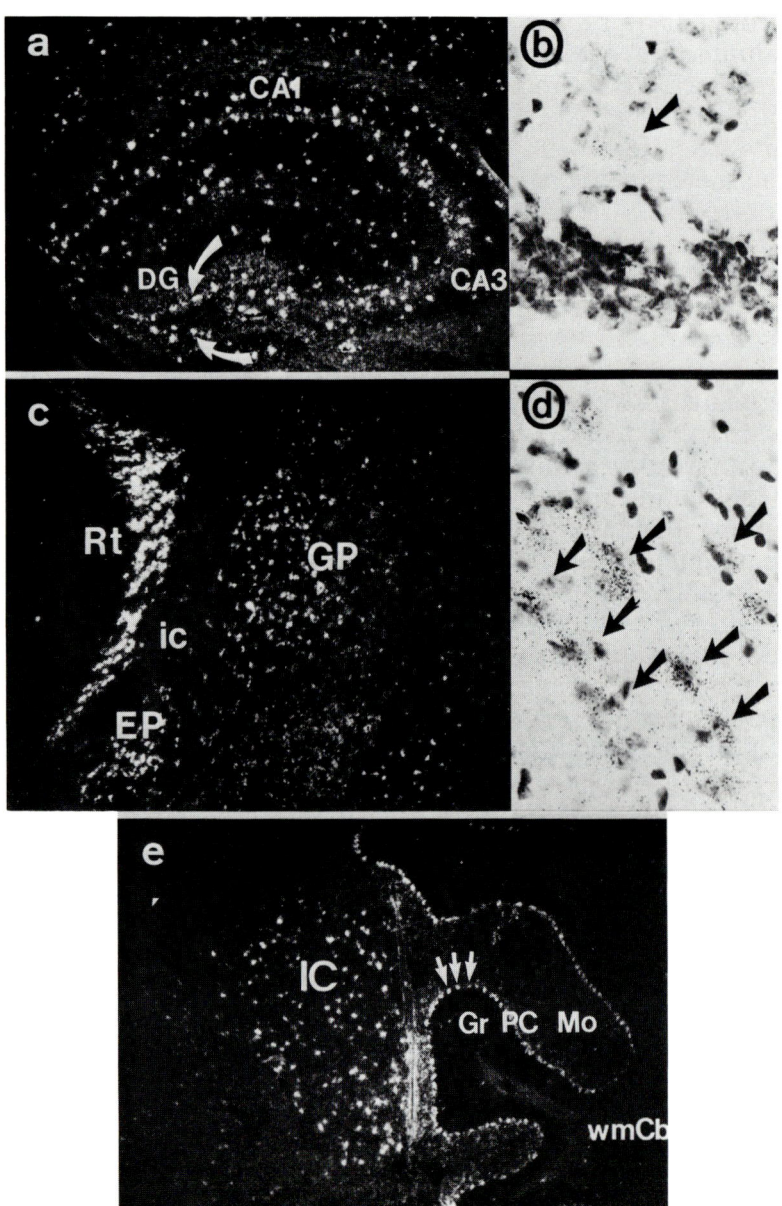

tion in combination with differential hybridization to map the CNS distribution of mRNAs whose expression is enriched in embryonic rat brain. *In situ* hybridization may also elucidate developmental events or reveal molecular characteristics that were not obvious using other approaches. For example, synthetic peptide-based immunocytochemical studies suggested that peptides derived from the brain-specific gene 1B236 may act in neurotransmission, in part, because they could be localized to specific subpopulations of neurons in the adult rat brain. However, *in situ* hybridization of 1B236 mRNA in the rat brain at early postnatal times revealed its presence in oligodendrocytes (Fig. 2A). It is now known that the sequence of 1B236 is identical to that of myelin-associated glycoprotein (Arquint et al., 1987; Lai et al., 1987), a known oligodendrocytic protein that may be involved in anchoring of the periaxonal myelin sheath (Quarles, 1984). In the adult rat brain, *in situ* hybridization for 1B236/MAG mRNA demonstrates its presence within certain neuronal populations (Fig. 2B), and suggests that it may act as cell-adhesion molecule for different cell types during development of the CNS.

←―――――――――――――――――――――――――――――――

Fig. 1. *In situ* hybridization to glutamate decarboxylase (GAD) mRNA detects the distribution of GABAergic neurons. A ^{35}S-labeled ssRNA antisense probe was synthesized from a plasmid template containing a 2.4-kb feline GAD cDNA sequence, generously provided by Allan J. Tobin (U.C.L.A.), with SP6 RNA polymerase and hybridized to mouse brain sections as described in Section 8. Panels a, c, and d are dark-field photomicrographs taken at a magnification of 40×; panels b and d, bright-field views at 200× after exposure to NTB-2 emulsion for 14 d. Panel a, coronal section of the hippocampus with arrows indicating GAD mRNA in the hilus bordering the dentate gyrus (FG). Panel b, hybridization to a single neuron flanking the granule cells of the DG. Panel c, coronal section indicating hybridization to neurons within the reticular thalamic nucleus (Rt), globus pallidus (GP), and entopeduncular nucleus (EP), but not within the internal capsule (ic). Panel d, single-cell resolution of GAD mRNA hybridization to neurons (arrows) within the Rt. Panel e, sagittal section showing hybridization to neurons of the inferior colliculus (IC) and Purkingie cells (PC) of the cerebellum (arrows); (Gr) granule cell layer, (Mo) molecular layer and (wmCb) white matter of the cerebellum (E. J. Hess and M. C. Wilson, unpublished).

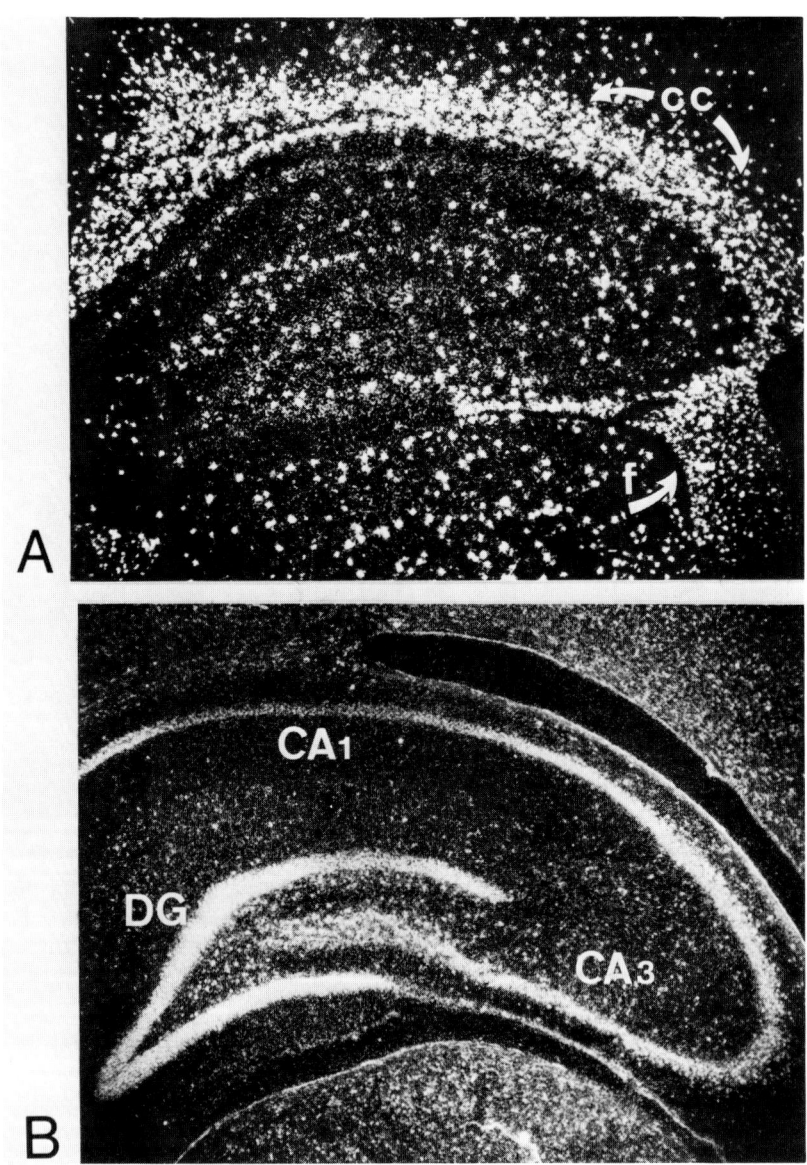

3. Kinetics of *In Situ* Hybridization

3.1. Stability of Hybrid Duplexes

The conditions for *in situ* hybridization are essentially those of conventional solution or filter hybridization with the important parameters of salt and formamide concentration, temperature, and probe G + C content and concentration. In solution, optimal rates of hybridization occur 20–25°C below the melting temperature (T_m) of the hybrid, a measure of the stability of the formed duplex (Wetmur and Davidson, 1968). The T_m of the duplex is dependent on the nature of the hybrid strands. RNA duplexes exhibit a 10–15°C higher T_m than corresponding DNA–DNA hybrids in aqueous solutions (Wetmur et al., 1981, Cox et al., 1984). Formamide is included within the hybridization solution, generally at 50% in 0.3–0.75M NaCl to destabilize the hybrids and thus depress the T_m. This allows the hybridization to be performed at temperatures of 45–55° where adequate tissue morphology is still retained. The reduction of T_m with formamide is significantly greater for DNA–DNA hybrids [0.65°C/% formamide (McConaughy et al., 1969) or DNA–RNA hybrids, 0.5°C/% formamide (Casey and Davidson, 1977)] than RNA duplexes [0.35°C/% formamide (Cox et al., 1984)]. In addition, the G + C content of the hybrid affects the T_m, since the greater hydrogen bond strength of GC base pairs serves to raise the T_m. For most practical purposes, an average G + C content of 40% may be assumed. Mismatch of base pairs de-

---

Fig. 2. Developmental shift in the cellular expression of myelin-associated glycoprotein (MAG) mRNA in the hippocampal formation of the rat. At postnatal d 20 (A), MAG mRNA is expressed in oligodendrocytes undergoing myelinogenesis. In the adult CNS (B), MAG mRNA is predominantly expressed by limited subpopulations of neurons. This study exemplifies the use of *in situ* hybridization to reveal sites of gene expression more sensitively than can be achieved with immunocytochemical studies (for details, refer to text). Dark-field photomicrographs of coronal sections hybridized with a single-stranded, ^{35}S-labeled RNA probe generated from the SP65 vector containing a cDNA homologous to the brain-specific mRNA 1B236 (Lai et al., 1987; Higgins et al., in preparation). Abbreviations: f, fornix; cc, corpus callosum; DG, dentate gyrus; CA1 and CA3, cornu Ammois fields of the hippocampus.

stabilizes the T_m of hybrids by approximately 1°C for each percent of unpaired bases in hybrids of 150 base pairs or more in length (Bonner et al., 1973). For hybrids greater than about 150 base pairs, the length of the probe has little effect on the T_m. The effect of these variables is summarized in the following equation for the thermal stability of DNA–DNA duplexes (Thomas and Dancis, 1973), which can be modified for the increased stability of RNA–DNA or RNA–RNA hybrids in aqueous solutions and the differential destabilizing affect of formamide for each hybrid.

$$T_m = \frac{81.5° \text{ C} + 16.61 \log M + 0.41 (\%GC) - 820 - 0.6° \text{ C}(\%F) - 1.4}{L} \text{ (\% mismatch)} \quad (1)$$

where 81.5°C is a constant for DNA–DNA hybrids, M is the molarity of ionic strength (mol/L), % GC is the G + C content as percent of total base pairs in hybrid, L is the length of the resultant hybrid, and 0.6°C (%F) is the effect of formamide destabilization of DNA duplexes multiplied by the percent formamide in the hybrid solution.

3.2. Stability of Hybrids Formed In Situ

RNA–RNA duplexes formed in *in situ* hybridization, however, exhibit a T_m 5°C lower than expected from solution hybridization (Brahic and Haase, 1978; Cox et al., 1984). This is probably the result of the preferential formation of hybrids under 150 base pairs in length, and is consistent with the observation that probes fragmented to 50 nucleotide lengths are optimal for *in situ* hybridization (Brahic and Haase, 1978; Angerer and Angerer, 1981). With oligonucleotides of less than 50 nucleotides, the stability of the hybrids formed is more significantly affected by their length, and consequently, more highly influenced by base pair mismatches and G + C content (Wallace et al., 1981).

3.3. Dextran Sulfate

The anionic polymer dextran sulfate is incorporated in the hybridization solution to promote higher hybridization efficiency. This is thought to result from an exclusion of the probe from the aqueous phase occupied by the dextran sulfate, and thus, increase the effective concentration of the probe and consequently its rate of hybridization (Wahl et al., 1979). With longer double-stranded DNA (dsDNA) probes, composed of complementary sequences, this effect of dextran sulfate is to generate extensive networks of reassociated sequences hybridized to the immobilized target RNA sequence and an amplified hybridization signal. Since most reports suggest that only short sequences, 200 nucleotides or less, participate in *in situ* hybridization to tissue sections, network formation of dsDNA probably does not occur or contribute greatly to the signal. However, amplification of *in situ* hybridization in cultured cells by probe networking promoted by dextran sulfate has been reported for dsDNA probe fragments >1500 nucleotides that contain both vector and probe sequence (Lawrence and Singer, 1985). The inclusion of dextran sulfate does result in a fivefold increase in hybridization with asymmetric single-stranded RNA (ssRNA) probes, probably by increasing the initial rate of hybrid formation (Angerer et al., 1987).

3.4. Rate of Hybridization

Assuming probe excess, hybridization to the RNA immobilized within the tissue ought to follow pseudo-first-order kinetics driven by the concentration of the probe as a function of time. A standard denotation of this relationship is the term $C_o t$, where C_o is the initial concentration of the probe in moles of nucleotide/L multiplied by the time in seconds. Probes of greater length or nucleotide complexity are, for the same $C_o t$ value, present at a lower molar concentration and, hence, drive the reaction at a slower rate. The reassociation rate of complementary strands of DNA in solution appears to be greater than the rate for sequences immobilized to filters (Flavell et al., 1974) or in tissue (Szabo et al., 1977). Therefore, with double-stranded probes, reassociation in solution competes with hybridization to the immobilized target sequence and eliminates the possibility of reaching hybrid saturation. The use of asymmetric single-stranded probes should circumvent this problem, and result in a quantitative saturation of the target dependent on probe concentration and time. However, the kinetics of *in situ* hybridization with single-stranded RNA (ssRNA) probes suggests that this does not occur and that the rate decreases after 4–6 h (Cox et al., 1984). Therefore, the extent of *in situ* hybridization appears primarily because of an initial rate as determined by a linear function of the probe concentration. The basis for the premature termination of hybridization is not clear, but it is unlikely that it is because of probe degradation (Cox et al., 1984), and may represent nonspecific binding of probe sequences or a decreased availability of target sequences during the hybridization. One solution is to employ a sufficiently high probe concentration to provide an initial hybridization rate that will saturate the target sequence before the hybridization ceases. Alternatively, probes of high specific activity, either ^{35}S- or ^{32}P-labeled ssRNA transcripts, may be used at a lower concentration to produce an equal, if not more intense, hybridization signal at subsaturation of the target sequence.

4. An Overview of *In Situ* Hybridization Methods

The major factors that determine the efficiency of *in situ* hybridization are the accessibility of the immobilized target sequence to the probe, which is largely affected by fixation procedures and subsequent pretreatment steps, and the selection of an appropriate probe. Since a variety of means can satisfy any one of the requirements for *in situ* hybridization, a number of methodological approaches have been developed. It is beyond the scope this article to describe fully the technical details of each method; however, an introduction to several important techniques may help in evaluating the appropriate method to study a particular system.

4.1. Preparation of the Tissue

For *in situ* hybridization, the target RNA sequence must be immobilized within the cellular matrix of the tissue section or cultured cell monolayer. Once fixed, the tissue or cellular matrix must be then permeabilized to expose the RNA to the probe by a series of pretreatments before hybridization. Basically, there are two methods for the initial fixation step: dehydration precipitation of RNA with ethanol:acetic acid, and chemical cross-linking of RNA and protein with aldehyde fixatives. For *in situ* hybridization to viral RNA sequences in cultured cells, ethanol:acetic acid fixation appears satisfactory and provides the sensitivity to detect viral sequences at very low abundance (Brahic and Haase, 1978), although postfixation with paraformaldehyde may improve hybridization (Haase et al., 1982). For cellular sequences, comparisons of aldehyde and ethanol:acetic acid fixation indicate that paraformaldehyde or glutaraldehyde cross-linking generally provides better retention of RNA and higher efficiency of hybridization (Godard and Jones, 1979; Angerer and Angerer, 1981; Gee and Roberts, 1983; Lawrence and Singer, 1985). The cells fixed with aldehydes, moreover, retain good morphology and are compatible with most immunocytochemical techniques. For most purposes of *in situ* hybridization, fixation of mammalian brain tissue is performed by perfusion of anesthetized laboratory animals or immersion fixation of dissected regions with 4% paraformaldehyde (Gee and Roberts, 1983), 0.5% formaldehyde and 0.5% glutaraldehyde, or 2% paraformaldehyde/0.075M lysine/0.01M NaIO$_4$ (Brahic et al., 1984). Satisfactory hybridization, however, can be obtained with fresh, frozen brain tissue if the sections are fixed with paraformaldehyde within several hours after sectioning (Branks and Wilson, 1986).

Both paraffin-embedded and frozen cut sections have been successfully used for *in situ* hybridization. Paraffin sections are generally thinner, 5–10 μm thick, and offer the potential to perform *in situ* hybridization (Lynn et al., 1983) or immunocytochemistry on serial sections of the same cell. However, for most purposes, thicker sections are cut with a cryostat microtome. For hybridization, sections are mounted on glass slides that have been coated (subbed) with protein or other substances to allow adherence. It is particularly critical to choose the correct subbing solution compatible with the prior fixation to avoid the disturbing loss of sections during hybridization and posthybridization washes. For example, slides subbed with acetylated Denhardt's solution (containing bovine serum albumin, polyvinylpyrrolidone, and ficoll) are well suited for sections of fresh, frozen brain (Branks and Wilson 1986), but will not reliably retain sections previously fixed with paraformaldehyde. For paraformaldehyde-fixed tissue, standard gelatin-chrome alum-coated slides are adequate. Other adhesives such as poly-L-lysine (Angerer et al., 1987), egg albumin, and Histostik (Brigati et al., 1983) have also been used with success.

4.2. Pretreatment

Fixation of the RNA by either precipitation with alcohol:acetic acid (3:1) and Carnoy's fixative, or cross-linking with aldehydes generally makes the tissue impenetrable to larger probes, although hybridization with short oligonucleotides probes appears not to be affected (Lewis et al., 1985). The extent of these pretreatments is a balance between completely disrupting the cytoplasmic matrix, causing poor cellular morphology and potential loss of the target RNA, and obtaining adequate and reproducible hybridization. It is possible that different cell types, for example oligodendrocytes and neurons, will differ in susceptibility to these treatments. Typically most protocols include detergent treatment with 0.01% Triton X-100 or 0.005% digitonin (Haase, 1987), further deproteinization with proteinase K at 1–50 μg/mL, and "etching" the tissue with 0.05–0.2N HCl. Between each treatment, the slides are usually washed once or twice by immersion in PBS (phosphate-buffered saline). It is important, however, that for any protocol adopted, especially with respect to the initial fixative, the investigator titrate these treatments with respect to cellular morphology and signal strength. For example, in the protocol used by our laboratory (described below), we have observed no effect of a detergent treatment of 0.01% Triton X-100, commonly used by other investigators. We have also tested the effect of HCl treatment that was originally incorporated to solubilize basic proteins (Harrison et al., 1973), but also reported to result in the loss of RNA (Godard and Jones, 1979, 1980). As shown in Fig. 3, we observed little difference in hybridization with a ^{35}S-labeled probe when the sections were treated with 0.2N HCl for 20 min (Brahic and Haase, 1978), 0.05N HCl for 7.5 min, or left untreated. It is possible, however, that HCl may specifically result in the loss of RNA from the surface of the tissue and seriously decrease the hybridization signal with ^{3}H probes because of the short emission path length of the isotope. We have also observed no significant difference between digestion with protease (Type IV, Sigma) at 0.25 mg/mL and proteinase K at 50 μg/mL. Proteinase K is preferable because it is a better characterized preparation and does not require predigestion to reduce residual nucleases.

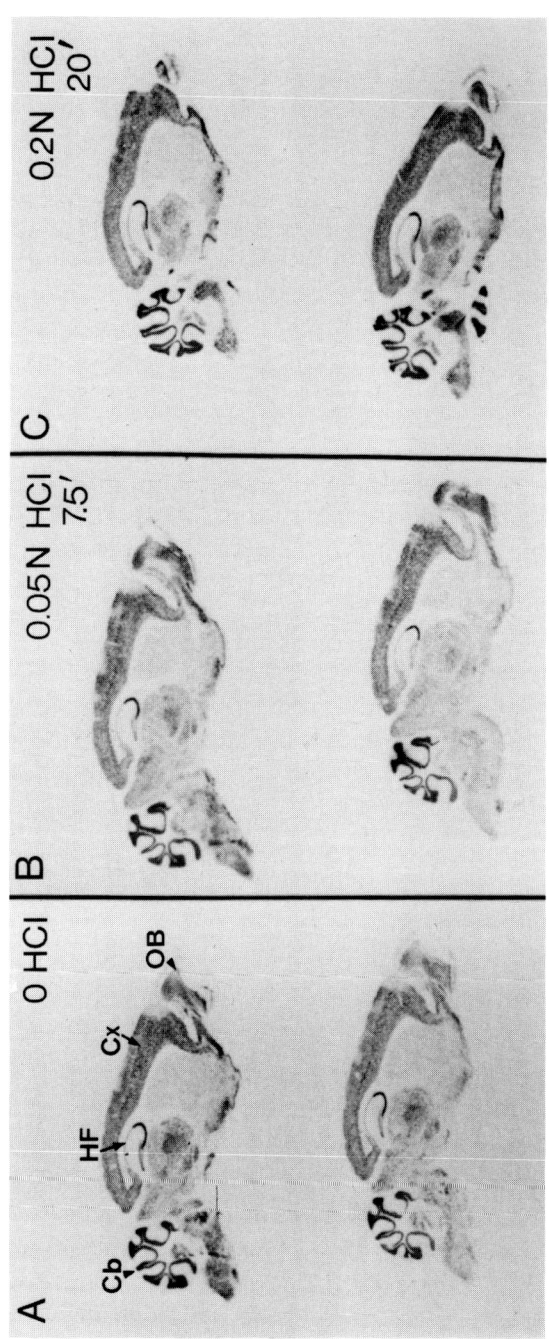

Fig. 3. The effect of HCl pretreatment on *in situ* hybridization. Sagittal sections of mouse brain were treated with HCl as indicated and hybridized with an ^{35}S ssRNA probe to a neuronal-specific mRNA, MuBr8 (Branks and Wilson, 1986) and exposed to Cronex 5 X-ray film for 15 h. Note that there is little significant difference in hybridization intensity between panels A, B, and C. Abbreviations: Cb, cerebellum; HF, hippocampal formation; Cx, cerebral cortex; OB, olfactory bulb. (R. Hart and M. C. Wilson, unpublished).

4.3. Probes for In Situ Hybridization

4.3.1. Double-Stranded DNA

The conventional method to prepare hybridization probes from recombinant DNA is to label double-stranded DNA (dsDNA) by "nick-translation" with *E. coli* DNA polymerase I (Rigby et al., 1977) or by random oligonucleotide primed synthesis with the Klenow fragment of Pol I, which lacks 5'–3' exonuclease activity of the holoenzyme (Feinberg and Vogelstein, 1983). To reduce background potentially contributed by labeled vector sequences, cDNA sequences are usually exised from the vector by cleavage with restriction endonucleases and gel purified before labeling. With ^{3}H-deoxynucleotide precursors a specific activity of up to 2×10^7 cpm/µg can be obtained (Gee and Roberts, 1983), whereas higher specific activities (>than 10^9) cpm/µg can be achieved with one or more ^{35}S- or ^{32}P-labeled nucleotides. Nick-translation of dsDNA incorporating ^{3}H- and ^{32}P-labeled nucleotides generally results in labeled fragments of 200–300 nucleotides in length because of the DNase I included in the reaction to induce the initial "nick" required for synthesis by DNA polymerase. Since short probes appear more efficient in penetrating the tissue, this length is satisfactory for *in situ* hybridization. However, when ^{35}S-labeled deoxynucleotides are incorporated, transcripts of higher molecular weight, approaching the full length of the cDNA template, are produced presumably because the synthesized DNA is refractory to subsequent DNase I cleavage. Size reduction of the labeled DNA can be accomplished by the increasing DNase I concentration 100-fold in the labeling reaction with no loss of specific activity (Branks and Wilson, 1986) or by limited acid depurination (unpublished observations). Alternatively, the template size may be reduced by cleavage with restriction endonucleases before labeling by the method of random oligonucleotide primed synthesis. The chief disadvantage of dsDNA is, since both strands of the probe are present in the hybridization reaction, the reassociation of the probe in solution competes with the hybridization to the immobilized RNA.

4.3.2. Single-Stranded DNA

Single-strand DNA probes can be synthesized from cDNA cloned in the single-stranded M13 DNA by priming with a synthetic oligonucleotide and synthesis with the Klenow fragment of *E.*

coli DNA polymerase (Akam, 1983). A discrete labeled transcript is produced by cleavage with a restriction endonuclease at a region made double-stranded in synthesis. The labeled DNA is then purified from the higher molecular weight vector DNA by denaturing polyacrylamide gel electrophoresis, and DNase digestion is used to reduce the length to 200 nucleotide fragments.

4.3.3. Single-Stranded RNA

Plasmids containing bacteriophage promoters can be used as templates for RNA polymerase to synthesize asymmetric single-stranded RNA (ssRNA) hybridization probes (Green et al., 1983). The adaptation and thorough characterization of the conditions for *in situ* hybridization with ssRNA probes has been explored by Angerer and colleagues (Cox et al., 1984). To construct the template, a cDNA sequence is inserted into a plasmid vector adjacent to either a SP6, T3, or T7 promoter sequence. This subcloning is facilitated by a polylinker sequence that provides multiple restriction endonuclease sites for ligation of the cDNA. Some vectors contain promoters for different RNA polymerases in opposing orientation, for example pGem (Promega Biotec) and pBS (Stratagene). Because of the promoter specificity of the RNA polymerases, labeled transcripts can be made from the same plasmid, either of the sense strand (the protein encoding strand of the mRNA) or of the antisense strand that is complementary to the mRNA. This can be useful, for example, when hybridization with a sense strand probe is used as a control for demonstrating the specificity of antisense hybridization. In ssRNA synthesis, ^{3}H-, ^{35}S-, and ^{32}P-labeled ribonucleotides can be incorporated, although ^{35}S-labeled nucleotides are incorporated with a lower efficiency. This adds further flexibility to the investigator with respect to the resolution and sensitivity afforded by the detection of these different labels. Moreover, since the probe is entirely synthetic, the specific activity of probes from various templates is essentially identical regardless of the efficiency of the reaction, assuming similar nucleotide compositions. In principle, this allows a direct comparison of hybridization signal intensities for different mRNAs when the lengths of the different probes are similar. This is an advantage over nick-translation labeling of dsDNA, in which the efficiency of the reaction can vary and result in probes of different specific activities because of the presence of "cold" template DNA in the probe preparation. The chief advantage, however, of ssRNA probes is that they are asymmetric, and therefore, the *in situ* hybridization is not compromised by competition with solution hybridization that occurs with dsDNA. The presence of equimolar complementary strands, for example, has been reported to decrease the signal of *in situ* hybridization as much as eightfold compared to that obtained with the single asymmetric RNA probe (Cox et al., 1984).

For *in situ* hybridization, the size of the RNA transcripts should be reduced to at least 150–200 nucleotides by alkali hydrolysis in 0.1–0.2N NaOH at 0°C or 40 mM NaHCO$_3$ 160 mM Na$_2$CO$_3$ pH 10.2 at 60°C (Cox et al, 1984). Since alkali treatment will also result in denaturation of the template DNA and introduce complementary sequences that will compete in solution hybridization, the DNA should be eliminated prior to hybridization. A control experiment to demonstrate the effectiveness of alkali hydrolysis is shown in Fig. 4. As seen in these panels, the greatest reduction in background is afforded by digestion of the template, with an increase in hybridization signal with alkali-broken transcripts.

A disadvantage of ssRNA probes is the relatively high concentration of ribonucleotide precursors needed to meet the apparent K_m values for SP6, T3, or T7 polymerases that range between 40 and 100 µM for ATP, CTP, and UTP (Chamberlin and Ryan, 1982). A biphasic requirement is observed for GTP and indicates a higher K_m value for the initiation of transcription with this nucleotide. Therefore, GTP should be present in excess and not isotopically labeled in generating ssRNA probes. When using ^{3}H-ribonucleoside triphosphates of low specific activity (generally 40–50 Ci/µmol), the incorporation of 100 µCi in a typical reaction volume of 25 µL results in a concentration of 64–80 µM, which approaches the apparent K_m for the substrate. However, in reactions utilizing high specific activity, the concentration of ^{35}S- or ^{32}P-labeled ribonucleoside triphosphates (800–3200 Ci/µmol), 250 µCi in a 25 µL reaction, is only 3–12.5 mM. This may be resolved either by supplementing with unlabeled substrate and decreasing the specific activity of the synthesized probe, or greatly increasing the amount of labeled ribonucleoside triphosphate, which can become quite expensive. Alternatively, the reaction may be performed below the K_m under limiting substrate concentrations. This results in an increase of prematurely terminated transcripts and therefore a higher proportion of the probe sequence complementary to the 3' end of mRNA.

RNA probes have been observed to be inherently more "sticky" and generate higher background signals than DNA probes. This is circumvented by posthybridization treatment with RNase A in 0.5M NaCl, which degrades nonspecifically bound probe but retains virtually all RNA-RNA hybrids. Although not routinely necessary, final post hybridization washes of high stringency, approaching the melting temperature (T_m) of the hybrids, can further reduce background. For example, stringent washes of 0.2 × SSC (30 mM NaCl, 3 mM sodium citrate) at 67°C or at 50°C with 50% formamide can be used, which in our hands does not result in significant damage to tissue morphology. Finally, high background has been generally observed with hybridization of ^{35}S-labeled RNA probes, presumably because of the formation of disulfide bonds between the probe and tissue. This is reduced by high concentrations of reducing agents (dithiothreitol, β-mercaptoethanol) in prehybridization, hybridization, and in an initial posthybridization wash.

4.3.4. Synthetic Oligonucleotide Probes

With the availability of automated DNA synthesizers, short oligonucleotide probes can be rapidly generated for *in situ* hybridization (Coghlan et al., 1985; Lewis et al., 1985). This permits the expedient use of published sequences in preparing probes without the need to resort to the isolation of cDNA clones or to subclone cDNAs into appropriate vectors for ssRNA probe synthesis. Synthetic oligodeoxynucleotides, as ssRNA probes, are single-stranded, and *in situ* hybridization is not subject to competition by solution reassociation of complementary sequences. They also have the advantage of DNA probes, being less "sticky." The short length (15–30 nucleotides) of most oligonucleotide probes allows for optimal penetration into the tissue. Pretreatment of the tissue, required for penetrance of longer cDNA and ssRNA fragments, therefore appears not to be necessary for the use of oligonucleotide probes in *in situ* hybridization (Coghlan et al., 1985; Lewis et al., 1986). Because of the short length of the oligonucleotide, the stability of hybrids formed is decreased and significantly affected by base composition of the probe (Wallace and Miyada, 1987). Hybridization with oligonucleotides, therefore, requires reduced temperature or increased ionic strength. For example, with oligonucleotide hybrids of 13–23 base pair lengths, the melting

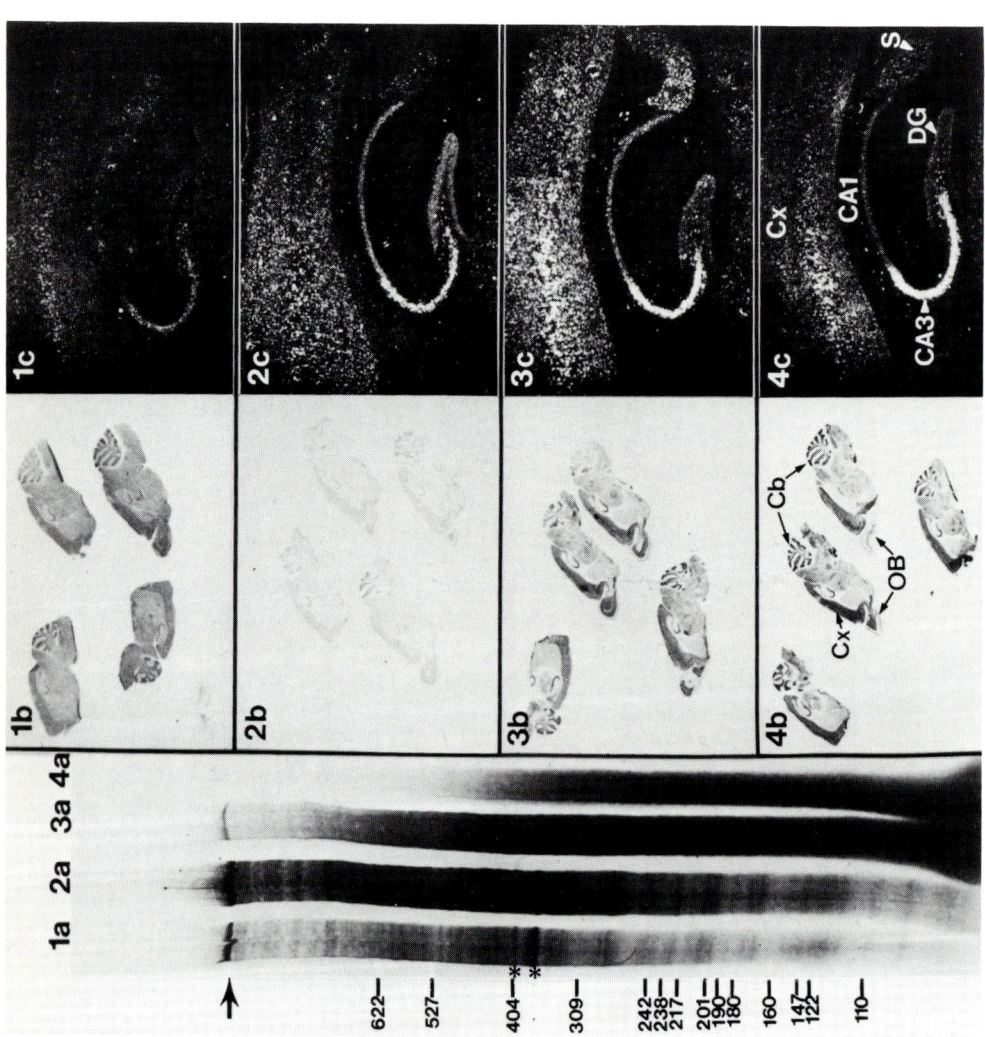

Fig. 4. The effect of DNase treatment and alkali breakage of ssRNA probe preparations. T7 RNA polymerase was used to generate a probe from a linearized plasmid template bearing a 881 cDNA nucleotide insert of neuron-specific MuBr8 mRNA. Lanes 1a–4a (left panel) show the size distribution of the transcripts on a denaturing urea polyacrylamide gel: Lane a, no treatment; lane b, after DNase I treatment at 400 U/mL for 30 min at 37°C; lanes c and d, after DNase and alkali treatment with 0.2N NaOH for 30 and 60 min, respectively (as described in Section 8.2.3.). Note the heterogenous length of the untreated transcripts resulting from the limiting concentration of CTP, 8.3 μM, in the synthesis reaction; the full length transcript is indicated by an arrow and the major prematurely terminated transcripts by asterisks. The middle panels, 1b–4b, show images of the hybridization with the probes indicated in lanes 1a–4a, respectively, after exposure to X-ray film for 16 h. The photomicrographs in the right panels, 1c–4c, show corresponding dark-field views of the hippocampus (40× magnification) from these sections after 4 d exposure to NTB-2 emulsion. Abbreviations: Cx, cerebral cortex; DG, dentate gyrus; S, subiculum; CA1 and CA3, cornu Ammonis fields of the hippocampus.

temperature in 1M NaCl, can be estimated with the formula: (T_m) of oligonucleotide hybrids, $T_m = 4(G + C) + 2(A + T)$ where G, C, A, and T are the number of corresponding nucleotides in the probe and the conditions for specific hybridization are met by reducing the temperature by 5°C (Wallace and Miyada, 1987). The effect of base composition, however, may not be predictable because of clustering of either weakly or strongly stable base pairs. For these reasons, the specific conditions for hybridization and posthybridization washes with a particular oligonucleotide should be established experimentally. It has been recently demonstrated that replacement of Na$^+$ by tetramethylammonium ions in posthybridization washes increases the stability of A/T base pairs to that of G/C base pairs in oligonucleotide hybrids (Wood et al., 1985). This greatly improves the precision for establishing conditions that distinguish faithful hybridization from nonspecific binding or mismatched hybridization because of related but nonidentical sequences. The use of this salt, however, has not been reported for *in situ* hybridization.

A potential disadvantage of oligonucleotide probes is a relative lack of sensitivity compared to more extensively labeled dsDNA or ssRNA probes. Although the oligonucleotide can be labeled to high specific activity either by efficient end-labeling with [γ-^{32}P] ATP by T4 polynucleotide kinase (Coghlan et al., 1985) or with addition of homopolymers of ^{3}H-, ^{35}S-, or ^{32}P-labeled deoxynucleotides by terminal transferase (Lewis et al., 1986), this still represents a maximum of 20–25 radiolabeled nucleotides per hybridized mRNA molecule, in contrast to 100 or more for moderate-length (i.e., 400 nucleotide) cDNAs uniformly labeled with a single nucleotide. Whether the enhanced penetrance of the oligonucleotide probe results in significantly more hybridization than randomly broken, but still diffusion-limited dsDNA or ssRNA probes, remains an open question until direct comparisons are made. However, it is evident that synthetic oligonucleotides are a suitable probe for *in situ* hybridization to at least more abundant mRNAs.

4.3.5. Biotinylated Probes

An alternative to the use of radioactive probes is the cytochemical detection of hybridization of probes labeled with biotinylated nucleotides (Langer et al., 1981). Biotinylated-11-dUTP is commercially available (Enzo Biochemical) and is incorporated into dsDNA probes by nick-translation with *E. coli* DNA polymerase I

(Langer et al., 1981). For *in situ* hybridization, detection is amplified by using antibiotin-IgG followed by a secondary anti-IgG antibody which is either conjugated with fluorochromes for indirect immunofluorescence or biotinylated for avidine-biotin peroxidase detection (Singer and Ward, 1982; Brigati et al., 1983). Although biotinylated probes have the advantage of eliminating exposure to radioactivity and offer an immediate detection of the hybridization, in contrast to sometimes lengthy autoradiographic exposures, it is difficult to quantitate the resultant signal and to determine its sensitivity. Studies with biotinylated probes, however, have been successful in detecting action mRNA present in less than 1000 copies/cell in cultured myoblasts by immunofluorescence (Singer and Ward, 1982) and viral sequences at a similar level of abundance with peroxidase enzymatic methods (Brigati et al., 1983). Biotinylated probes, moreover, promise a high level of resolution that has been used to demonstrate the nonrandom distribution of cytoskeletal protein mRNAs (Lawrence and Singer, 1986). Recently, the chromosomal location of a single-copy gene on metaphase chromatids and in nuclei at interphase was reported using a biotinylated probe with fluorescein avidin as the detection reagent (Lawrence et al., 1988).

5. Controls for *In Situ* Hybridization

Although there is no single experiment to distinguish all possible artifacts arising from *in situ* hybridization, several different parameters can be examined, which together can verify the specificity of hybridization. The objective of most controls is to distinguish specific hybridization from adventitious, nonspecific adherence of the probe sequence within the tissue. For example, we have observed that the most common pattern of nonspecific binding is dependent on cell density, similar to a Nissl stain, with background particularly high in the granule cell layer of the cerebellum and dentate gyrus of the hippocampus. In some cases, nonspecific binding to myleinated tracts has also been observed. Consequently, when initially optimizing the parameters of *in situ* hybridization, it is suggested that a probe to a mRNA expressed in a localized and well-defined group of cells is used rather than a probe to a highly distributed mRNA such as actin mRNA.

5.1. RNase Pretreatment

The simplest demonstration of the specificity of hybridization is to treat alternate sections with ribonuclease before hybridization to remove the specific target sequence as well as most other RNA species. This is conveniently done after pretreatment with proteinase K and HCl, but prior to the final paraformaldehyde fixation (described in the protocol below) by incubating selected slides in $0.3M$ NaCl, 10 mM Tris pH 7.5 and 1 mM EDTA (or 2× SSC) containing 50 µg/mL RNase A at 37°C. The nuclease is then inactivated by the final postfixation with 4% paraformaldehyde.

5.2. Heterologous Probes

To evaluate nonspecific hybridization to total RNA, possibly resulting from low stringency in hybridization conditions, the use of heterologous probes to demonstrate distinct patterns of hybridization is valuable. For example, as shown in Fig. 5, a dsDNA probe to growth hormone mRNA results in a strong signal over the anterior lobe of the pituitary, which was embedded in liver tissue before sectioning, but not to adjacent brain tissue. The cDNA MuBr85, on the other hand, shows a weaker signal to pituitary, but not to the surrounding liver tissue, and considerably more hybridization to selected brain regions. Neither probe generates a signal to either pituitary or brain sections after RNase treatment. With asymmetric ssRNA probes, particularly those generated from vectors containing opposing RNA promoters, the pattern obtained with the antisense probe can be compared with that of sense strand transcripts. Although sense and antisense probes ought to be of similar length and identical specific activity, they do not share any sequence similarity with the exception of inverted repeats and should, therefore, be considered heterologous probes. Hybridization of sense-strand probes, however, will determine if the probe sequence has regions homologous to repetitive elements present in genomic DNA that might obscure mRNA hybridization with the antisense probe.

5.3. Other Controls of Hybridization Specificity

Consistent patterns of hybridization with probes to different regions of the mRNA provide an additional control for specificity. This is most applicable with oligonucleotide probes made to adja-

In Situ Hybridization

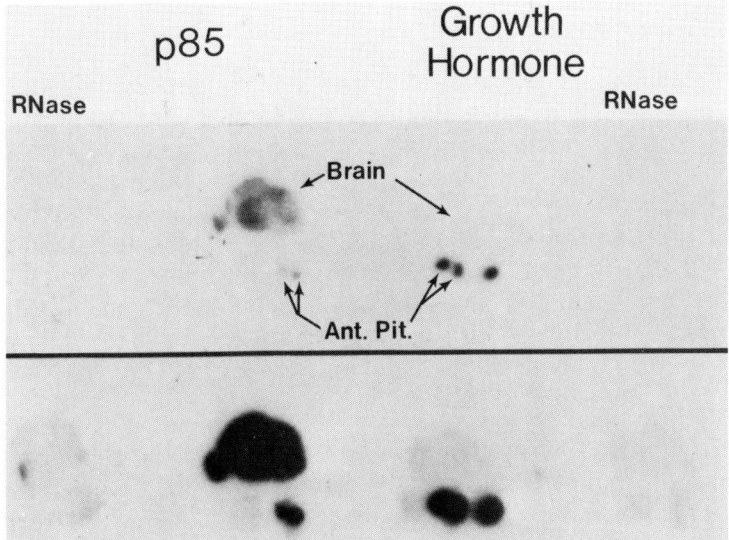

Fig. 5. Control for the specificity of *in situ* hybridization. cDNA inserts corresponding to a brain expressed mRNA, MuBr85 (Branks and Wilson, 1986) and growth hormone mRNA (a generous gift of R. Evans, Salk Institute) were labeled with ^{32}P by nick-translation and hybridized to sections of fresh, frozen mouse brain and of pituitary embedded in liver tissue for sectioning. The sections were fixed with 4% paraformaldehyde after mounting on slides coated with acetylated Denhardt's solution. To the left and right are comparable sections treated with RNase A prior to hybridization. The upper panel shows the hybridization after an overnight exposure to X-ray film. The lower panel is an overexposure to demonstrate the low background levels of nonspecific probe adherence of the growth hormone probe to brain and of both probes to liver tissue. Specific hybridization to the anterior pituitary (Ant. Pit.) with both probes is indicated.

cent antisense sequences (Uhl et al., 1985). It must be cautioned, however, that alternative splicing of the mRNA precursors may result in different patterns of expression among different cell populations (Amara et al., 1982). The specificity of hybridization can also be determined by a comparison of the T_m of hybrids formed in solution and *in situ*. The thermal stability of a duplex formed from homologous complementary RNA sequences *in situ*, for example, should be 5°C reduced from those formed in solution under similar conditions (Brahic and Haase 1978; Cox et al., 1984).

Other control experiments can be considered that may be, in some cases, less successful for demonstrating specific hybridization. For example, attempts to block specific hybridization by preincubation with an excess of cold probe sequence may be incomplete, reducing the hybridization by only 80–90% of unblocked sections (Gee and Roberts, 1983). This may reflect "uncovering" of targets during subsequent hybridization incubation. Comparison of the coincidence of *in situ* hybridization to the mRNA and immunocytochemical localization of the protein has been used to confirm *in situ* hybridization results (Gee and Roberts, 1983; Brahic et al., 1984; Shivers et al, (1986). However, translational control, as well as rapid degradation, or transport and delivery of the antigen to other cells may lead to conflicting views by hybridization and immunological techniques. For example, in the Brattleboro rat, a point mutation within the vasopressin sequence precludes translation of vasopressin (Schmale and Richter, 1984), but does not greatly affect the expression of the vasopressin-neurophysin precursor mRNA (Uhl et al., 1985).

6. Simultaneous Detection of Multiple-Gene Products

6.1. *Immunocytochemistry and* In Situ *Hybridization*

The combination of *in situ* hybridization and immunocytochemistry has been used to detect viral antigens (Brahic et al., 1984) and neuropeptides (Shivers et al., 1986) and their respective mRNAs in the same cell, as well as to demonstrate the coexpression of different neuropeptides within the same cell (Wolfson et al., 1985). Because of the extensive treatments required for both techniques, there is some loss of sensitivity to either detection method (Brahic et al., 1984), suggesting that this may be limited to evaluating gene products of higher abundance. Since the prehybridization treatments for *in situ* hybridization are likely to destroy the epitopes recognized by most antisera, immunocytochemistry is carried out first. This is typically done using a standard secondary biotinylated anti-IgG coupled to avidin biotinylated horseradish peroxidase complex and detection with diaminobenzidine tetrahydrochloride (DAB) as the chromogen. The resulting complex is insoluble and survives the hybridization treatments. For compatibility with immunocytochemistry and *in situ* hybridization,

In Situ Hybridization

tissues are fixed by perfusion with paraformaldehyde alone (Wolfson et al., 1985; Shivers et al., 1986), or 0.5% paraformaldehyde and 0.5% glutaraldehyde (Brahic et al., 1984). Both paraffin-mounted sections (Brahic et al., 1984) and cryostat-cut, floating sections (Wolfson et al., 1985) have been used. Degradation of RNA in the section by RNase in the antisera can be prevented by including diethyl pyrocarbonate during immunocytochemistry (Shivers et al., 1986). To block extensive nonspecific binding of RNA probes to DAB complexes, the sections are acetylated with acetic anhydride (0.25%) in O.1M triethanolamine (pH 7.5) before hybridization (Brahic et al., 1984).

6.2. Double-Label Hybridization

Double-labeling with different probes labeled with ^{3}H and ^{35}S has been recently reported (Haase et al., 1985). The two isotopes are distinguished by development of colored grains in two layers of emulsion. Although not in use in many laboratories, this technique presents the opportunity to quantitate the relative levels of mRNA expressed by two genes within the same cell, a potential not applicable with qualitative immunocytochemistry.

7. Quantitation of In Situ Hybridization: Problems and Potential

A critical question in neurobiology is whether *in situ* hybridization can be used as a quantitative tool to determine the absolute amount of a given mRNA and therefore compare relative levels of gene expression between neuronal populations under different physiological or behavioral conditions. Under hybridization conditions where the probe is in vast excess, the abundance of target mRNA does not influence the rate of hybridization. Thus, the intensity of the resultant signal is an accurate representation of the abundance of the target RNA, even if the kinetics of the reaction are not ideal pseudo-first order in nature (Cox et al., 1984). When true saturation of the target is reached under high probe concentrations, it may be possible to calculate directly the number of mRNA copies per cell based upon extrapolations from probe-specific activity, probe length, and efficiency of autoradiographic grain development (Brahic and Haase, 1978; Cox et al., 1984; Rogers, 1979). However, even when such conditions cannot be met, for example, when using small amounts of high specific activity ^{35}S-labeled RNA probes, these probes can still be used to measure differences in the relative abundance of mRNAs between cell populations (Young, 1986).

Many technical issues concerning quantitative autoradiography, and its application to nervous system tissue, have been addressed in ligand-receptor binding studies (Kuhar and Unnerstall, 1985). One problem for quantitation is that the density of grain development with emulsion or X-ray film may not be a linear reflection of radioactive emission within a tissue section. Thus, at low signal intensities, optical density may increase rapidly as a function of radioactive emission, and at higher intensities, the responsiveness of the emulsion may steadily decrease until saturation is reached. For example, Ehn and Larsson (1979) plotted the relationship between optical density in ^{3}H-Ultrofilm (LKB Industries) and varying concentrations of tritium in brain paste standards, and found that the film had a linear response range over a small range of optical densities. Another problem, which arises when using low energy β-emitters such as tritium, is that different brain regions, based on the mixture of grey and white matter, may vary in density and thus will differentially quench tritium decay (Kuhar and Unnerstall, 1985). Additionally, variability in emulsion thickness may occur when slides are dipped in liquid emulsion. Grain density may reflect emulsion thickness with a high energy isotope such as ^{35}S, leading to possible errors in quantitation by grain counting or optical density analysis. More even emulsion layers can be obtained with film or by use of dry emulsion-coated coverslips (Young, 1986).

7.1. Densometric Analysis of X-Ray Film Images

Given the caveats to quantitation of autoradiography, computer-enhanced image analysis of X-ray film autoradiographs promises to be useful for analyzing patterns of gene expression detected by *in situ* hybridization. For X-ray film images of hybrids labeled with higher β-emission isotopes, ^{32}P and ^{35}S, computer-assisted digitalization of the intensity of the autoradiographic signal is measured as optical density, and results in a color-coded measurement of the relative signal strength. To extend the quantitation beyond the linear response of the film, comparisons can be made with calibrated standards of relative radioactivity coexposed with the hybridized section. The autoradiographic signal may then be expressed in terms of radioactivity and permit more direct comparisons between individual experiments, regardless of length of exposure to film. Storage of digital images also allows for the comparison of signals obtained with adjacent sections that may be hybridized in parallel to heterologous probes. For example, background signals resulting from nonspecific binding can be evaluated by scanning sections exposed to sense strand probes and subtracted from antisense probe hybridization when ssRNA transcripts are used. Similarly, it may be useful to compare the abundance and regulated expression of a specific mRNA with respect to a "housekeeping" gene, constitutively expressed and thus representative of the total mRNA in all cell populations.

7.1.1. Three-Dimensional Reconstructions

Perhaps most intriguing is the potential to compile of a series of two-dimensional images into a three-dimensional (3-D) reconstruction, an approach first applied to the neurosciences by Levinthal and Ware (1972). Considering the complex cytoarchitectonic organization of the brain, it is likely that, by generating a 3-D image of *in situ* hybridization patterns, further insights can be gained into the identification of structures having shared gene expression. To investigate this possibility, in collaboration with R. A. Hawkins at the Pennsylvania State University College of Medicine, a preliminary study was made to prepare a 3-D reconstruction of *in situ* hybridization (Fig. 6) of a ^{35}S-labeled ssRNA probe to a neuronal-specific mRNA, MuBr8 (Branks and Wilson, 1986; Oyler et al., in preparation). Autoradiographs of the hybridization to a series of coronal sections were scanned and digitized at 100 µm resolution with a computer-operated microdensitometer (Optronics P-1000) and each pixel converted into optical density valves. Conventionally, alignment of the sections is accomplished by interactive manipulation of the images; however, in this case, the alignment was executed entirely by a program developed by Hibbard et al. (1987), run on a DEC VAX 11/780 computer. After the images of coronal sections were edited for artifacts in the surrounding field, this program determined the center of mass for each image and established the principal axes through the center of each section. The set of images were subsequently aligned by a translation of the principal axes to a common origin of reference and

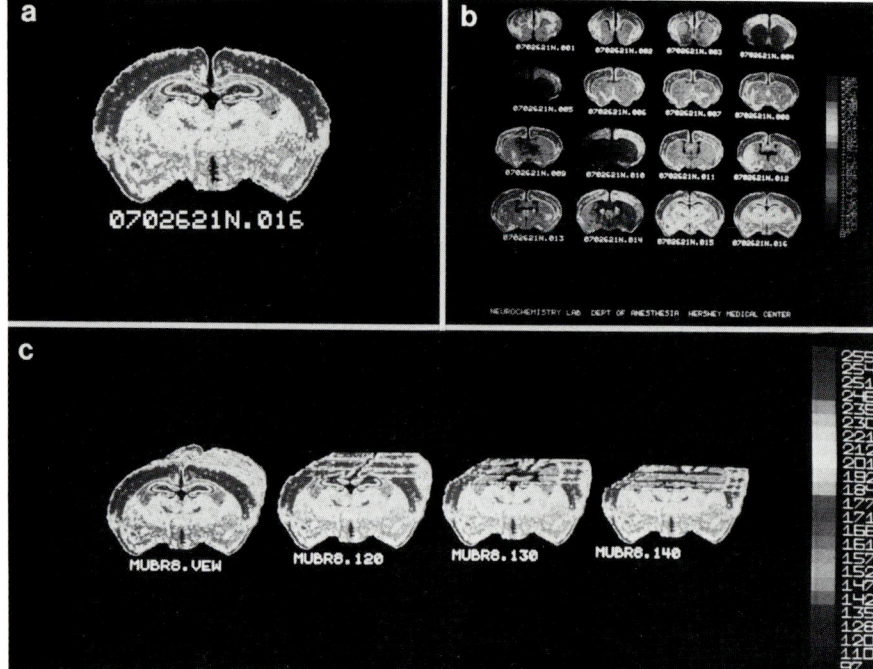

Fig. 6. Color-enhanced images (reproduced in black and white) of MuBr8 *in situ* hybridization in coronal sections and as a 3D reconstruction of the brain. Autoradiographs of 16 coronal brain sections hybridized with a probe to MuBr8 were digitized using a computer-driven optic densitometer and depicted as color images on an AED-1024 terminal. A representative coronal image is shown in Panel A. The complete data set of 16 coronal sections is presented in Panel B. The coronal images were aligned and assembled to reconstruct a 3D image of the brain in Panel C. (G. A. Oyler and M. C. Wilson, unpublished).

rotated to place the longer principal axis horizontal. The reconstructions may then be viewed as stereograms and displayed at several horizontal planes (panel c). Although this example lacks the resolution obtained with glucose metabolism studies (Hibbard et al., 1987), possibly because of the tissue treatments required for *in situ* hybridization, it does provide information not appreciated in viewing the primary autoradiograph. For example, the laminar distribution in the neocortex and diminished intensity in ventral cortical regions is more apparent in the 3-D reconstruction.

7.2. Grain Counts at the Cellular Level

At the cellular level, quantitation of in situ hybridization is best achieved with ^{3}H- and ^{35}S-labeled probes. In general, slides are coated with Kodak NTB-2 emulsion; however, exposure to Kodak NTB-3 or Ilford L4 nuclear emulsion results in finer grains and may provide higher resolution. For NTB-2, the efficiency in developing a latent image is 0.1 and 0.5 grains per ^{3}H- and ^{35}S-disintegration, respectively (Rogers, 1979). The most common method to determine autoradiographic grain density is by manual microscopic examination. Obviously, this is a laborious and tedious task with ample room for subjective error. Moreover, counting grain densities of ^{35}S-labeled probe hybridizations is difficult because of the development of grains throughout the emulsion and in varying focal planes of the microscopic image. Under bright-field illumination, grain counts of ^{3}H-probe hybridization are probably more reliable because of its less energetic β-emission and, therefore, shorter track length.

Recently, several commercially available video-camera-linked personal computer systems have been developed that, in concert with other morphological measurements, are designed for automated grain counts (for example, Nikon/Joyce-Loebl Magiscan, Olympus Cue, Amersham RAS, and Imaging Research MCID systems). With these systems, the operator circumscribes the area of the cell, and the number of targets is counted and related to the target area as well as the total area of the microscopic field. In collaboration with other investigators of Scripps Clinic and Research Foundation, we have employed the computer-assisted neurocytology system EMMA, an acronym for Electronic Morphometry and Mapping Analysis (Young et al., 1985), to quantitate in situ hybridization. This system also has the capacity for synthetic reconstruction of the images with modeling programs and integration of the empirical data into standardized data bases such as brain atlases and reference data bases, and operates on a DEC Microvax II UNIX-based computer system with an image processing system, 1024XM, from Megavision (Santa Barbara, CA). For autoradiographic grain counts, tissue sections are imaged under dark-field illumination at a magnification of ×200 with a resolution of 1024 × 1024 pixels of 256 gray scale values, and grain clusters are taken to represent individual cells. When isolated from background, the information of the entire field is related as optical density per cell with variance computed as standard deviation. Excessive autoradiographic exposure is avoided, which would result in an underestimation of the grain density because of oversaturation of the image. In a recent study, this system was used to make comparisons of amyloid-β-protein mRNA levels in individual hippocampal neurons of normal aged and Alzheimer's disease brain tissue, demonstrating a two–threefold increased expression of the mRNA in parasubicular neurons relative to CA3 pyramidal neurons in Alzheimer's diseased brain (Higgins et al., 1988).

8. A Procedure for *In Situ* Hybridization

The protocol for *in situ* hybridization currently used by this laboratory has been devised to provide an optimal signal to background ratio while preserving tissue and cellular morphology. We have also made an effort to prepare the tissue in a manner compatible with standard immunocytochemical and histochemical reagents. If both immunocytochemical and *in situ* hybridization is to be attempted within the same section, immunocytochemistry must precede the prehybridization treatments that include exposure to proteinase K and HCl. Ribonuclease treatment following the hybridization is used to eliminate the background of nonspecific bound probe, but also results in the degradation of cytoplasmic RNA. Since this eliminates most of the cytoplasmic stain by cresyl violet or hematoxylin/eosin, although nuclear staining is retained, alternate sections should be processed separately for histochemical analysis.

The procedure is outlined for the use of ^{35}S-asymmetric ssRNA probes, which allows for X-ray film and emulsion autoradiography. The procedure is compatible with dsDNA probes labeled with ^{3}H or ^{32}P by nick-translation or random oligonucleotide primers; however, in labeling with ^{35}S by nick-translation, further measures to reduce the size of the probe are required (*see above* and Branks and Wilson, 1986).

8.1. Preparation of Tissue

Brains of laboratory animals are fixed by transcardiac perfusion with freshly prepared 4% paraformaldehyde in isotonic,

In Situ Hybridization

0.15M sodium phosphate buffer, pH 7.4 (PB). Human autopsy specimens are best obtained within hours of death, and dissected regions are fixed by immersion in 4% paraformaldehyde/PB solution for 48 h. The tissue is then passed through graded sucrose solutions for cryoprotection. In the preparation of the 4% paraformaldehyde/PB, it is critical to avoid temperatures exceeding 75–80°C to prevent substantial depolymerization of the paraformaldehyde.

8.1.1. Perfusion Fixation

Mice or rats are anesthetized with CO_2 or by administration of chloral hydrate (400 mg/kg, ip) for mice or pentobarbital (40 mg/kg, ip) for rats until there is no motor reflex. The rib cage is then opened to expose the heart, and a cut is made in the right atrium to provide an exit for the fixative. An intravenous catheter (No. 14 angiocath), connected to the 4% paraformaldehyde/PB solution, is inserted into the left ventricle, and a clamp is slowly opened after removal of the needle. The fixative is allowed to flow into the heart by a gravity feed of about 1.5 m for 3–5 min, until the body is rigid, eyes clear, and the liver becomes pinkish. The brain is removed intact and placed in a sterile 50-mL plastic centrifuge tube containing 4% paraformaldehyde/PB for 4 h at 4°C. For cyroprotection, the fixative is decanted and replaced with cold graded sucrose solutions of 12, 16, and 18% in 0.15M phosphate-buffered saline (PBS) or directly stored overnight in 18% sucrose/PBS at 4°C. Equilibrium with the final 18% sucrose solution is indicated by the brain sinking to the bottom of the tube.

Preparation of brain tissue from primates, for example of cynomologus monkey (Macaca fascicularis), requires a more extensive perfusion technique. The animals are deeply anesthetized with ketamine hydrochloride (25 mg/kg of bodyweight, intramuscularly) and sodium pentobarbital (10 mg/kg, ip). Mechanical ventilation is performed with 100% oxygen via an endotracheal tube. The chest is opened to expose the heart, and 1.5–2 mL of a 1% aqueous sodium nitrate is injected into the left ventricle. Animals are then perfused transcardially, cutting the right atrium and inserting a cannula into the left ventricle, first with cold 1% paraformaldehyde/PB for 30–60 s followed by 4% paraformaldehyde/PB for 6–8 min at a flow rate of 250–500 mL/min with a peristaltic pump. The brain is removed, cut into blocks 3–5 mm on a side, and passed through graded sucrose solutions for cyroprotection as described above.

8.1.2. Cryomicrotomy

Intact rodent brains or primate tissue blocks are fixed to chucks with O. C. T. compound (Tissue Tek) on a freezing stage or dry ice. Ten–20-μm sections are collected directly on gelatin-chrom alum-subbed slides and air dried. As a safeguard against RNase degradation, gloves are worn when handling the tissue or slides. In order to control for hybridization variability between slides, we routinely attempt to place adjacent sections on serial slides in a "round robin" manner. Similarly experimental and control brain sections are placed on the same slide, thus by necessity requiring them to remain at room temperature for the duration of the cutting. Usually four sections are placed on a slide in an area covered by a 22 × 50 mm coverslip. In our experience, sections are good for 2–3 d at 4°C in a sealed box or up to 4 wk at –70°C. We recently observed, however, that 6-wk-old sections kept dry at room temperature in a sealed slide box were indistinguishable after hybridization from sections cut just prior to use. The loss of hybridization signal with stored slide mounted sections, therefore, may reflect rehydration of the tissue.

8.1.3. Gelatin-Chrom Alum-Subbed Slides

On occasion, we have observed that there is variability in an retaining sections through the posthybridization washes with gelatin-chrome alum-subbed prepared weeks before use. This can be avoided by using slides subbed the day before use. A standard gelatin-chrom alum-subbing solution is prepared by dissolving 4 g of gelatin (bovine skin, type III, 225 Bloom, Sigma) in 400 mL distilled H_2O with stirring at 37°C. When completely dissolved, 5 mL of 20 mg/mL $CrKSO_4$ solution in distilled H_2O is added, and the solution is allowed to sit in a glass staining dish at 37°C for 30–60 min. Precleaned microscope slides are immersed in the gelatin-chrom alum solution for approximately 10 s in a slide rack and allowed to dry overnight at RT.

8.2. Preparation of ssRNA Probes

8.2.1. Template

To obtain optimal in vitro transcription, plasmid DNA preparations are extensively purified, preferably through CsCl/ethidium bromide equilibrium density gradient fractionation (Maniatis et al., 1982). Before transcription, the plasmid template must be linearized and care must be taken that the digestion goes to completion, as supercoiled plasmids are preferential templates for the bacteriophage RNA polymerases. Digestion is done with an appropriate restriction endonuclease in excess chosen to cleave, preferentially once, within the polylinker sequence as close to the 5' end of the mRNA insert for antisense probes or at the 3' end for generation of sense strand transcripts. The enzyme should leave either a 5' overhang sequence or a blunt end. The short 3' single strand left by some enzymes can serve as polymerase initiation sites to generate a complementary strand. It is also preferable to use plasmids with inserts lacking homopolymer sequences, often generated by homopolymeric tailing during cDNA cloning, since these sequences appear to induce premature termination of transcription. Complete digestion is monitored by comparing an aliquot of digested and untreated plasmid after electrophoretic fractionation on 1% agarose gels stained with ethidium bromide (Maniatis et al., 1982). If completely cleaved, the digested DNA will migrate as a single, homogenous band behind the supercoiled starting material. A typical digestion is:

10 µg CsCl purified plasmid DNA
5 µL 10× Buffer (as indicated by the supplier)
5 µL 1 mg/mL Bovine serum albumin (DNase free)
H_2O to 45 µL
5 µL Restriction endonuclease (5–10 U/µL)
Incubation at 37°C for 3–5 h

After determining that cleavage has gone to completion, the digest is diluted to 200 µL with 0.2M sodium acetate, pH 5.2, 10 mM EDTA, and extracted twice with an equal vol of phenol:chloroform:isoamyl alcohol (25:24:1) saturated with the 0.1M sodium acetate, 10 mM EDTA, followed by an extraction with 1 vol of chloroform:isoamyl alcohol (24:1) and precipitation with 2 vol of ethanol at –20°C. The plasmid is collected by centrifugation in a microfuge for 15 min at 4°C, washed once with 70% ethanol containing 1 mM EDTA, and dried before resuspending at 1 µg/µL.

8.2.2. Transcription Reaction

In synthesis of high specific-activity probes of 2×10^9 dpm/µg, the transcription reaction generates 50–120 ng of transcript. If more probe is desired, increased amounts of ^{35}S-CTP of lower specific activity are used. For example, 0.5 mCi of ^{35}S-CTP at 800 Ci/μmol results in 500–600 ng of transcript with a specific activity of 1.3×10^9 dpm/μg. Incorporation of ^{3}H-ribonucleoside triphosphates yields significantly more transcripts with specific activities of 0.6–2×10^8 dpm/μg, depending on how many labeled precursors are used.

    1–1.5 μg of linearized plasmid template
    5 μL 5x enzyme buffer (as recommended by the supplier)
    1 μL RNasin (800–1000 U Promega Biotec, or RNase Block, Stratagene)
    1 μL 50 mM DTT
    1 μL 10 mM ATP
    1 μL 10 mM GTP
    1 μL 10 mM UTP
    4 μL ^{35}S-CTP (1200–1400 Ci/μol, 50–60 mCi/mL, NEN)
H$_2$O to 23.5 μL
    1.5 μL RNA polymerase (50 U/μL)
    Incubation at 37°C for 1.5–2 h

Synthesis is measured as percent incorporation (usually 40–50%) by removing a 1 μL sample and diluting it in 110 μL of H$_2$O containing 50 μg of carrier yeast RNA. Ten μl is spotted directly on a filter, and the remaining 100 μL is precipitated with 20% TCA, collected on a nitrocellulose filter dried, and counted with scintillation fluid. Assuming that one of four polymerized nucleotides is a labeled CTP residue, the following equation can be used to estimate the mass of transcript produced.

$$\text{ng probe} = \frac{\text{incorporation} \times \text{input} \, ^{35}\text{S-CTP} \times 340 \times 10^6 \text{ng/μmol} \times 4}{\text{sp. act. of CTP}} \quad (2)$$

where:
    Incorporation is the fraction of the label TCA precipitable
    Input is the amount of label in the reaction given as Ci
    The specific radioactivity of the CTP is given as Ci/μMol
    340 is the average mol wt of ribonucleotides
    4 is the number of ribonucleotides incorporated.

8.2.3. Purification of the Labeled Probe

In order to provide optimal signal-to-noise ratio, a rapid procedure is outlined to remove the template DNA by DNase I digestion and for alkali breakge of the probe RNA transcripts to allow penetrance of the probe into the tissue section.

24 µL transcription reaction
1 µL RNase-free DNase I (10,000 U/mL, Stratagene)
   Incubation at 37°C, 30 min
60 µL 10 mM EDTA, 10 mM Tris pH 7.4, 0.2% SDS
1.67 µL 5M NaCl
1 µL 1M DTT
Equilibrate to 0°C on ice
10 µL 2N NaOH
Incubation at 0°C, 45 min.
20 µL 2M HEPES (not buffered, for neutralization)

The solution is then extracted once with phenol:chloroform:isoamyl alcohol (25:24:1) saturated with 0.1M sodium acetate, 10 mM EDTA, and the aqueous phase is applied directly to a G50 Sephadex "Nick column" (Pharmacia) preequilibrated with 10 mM EDTA, 10 mM Tris pH 7.4, 10 mM DTT. The bed vol is eluted in 0.4 mL of the equilibrium buffer, and the RNA transcripts are collected from a second 0.4-mL wash in a sterile 1.5-mL microfuge tube. Sodium acetate, pH 5.2, is added to 0.2M and the RNA is precipitated with 2 vol of ethanol (1 mL) in a dry ice ethanol bath or overnight at –20°C. Approximately 95% of the labeled transcripts are recovered.

8.3. Prehybridization Treatments

It is convenient to prepare both pre- and posthybridization treatments in glass staining dishes containing 500 mL of the indicated solutions and progress through with slides held in stainless-steel slide racks. The solutions are made fresh with H_2O treated with 0.1% diethylpyrocarbonate and autoclaved. The staining dishes are stored in 10% nitric acid and rinsed in distilled H_2O prior to use. The dish containing RNase is, however, sequestered and not mixed with the others. The pretreatment steps include:

1. Postfixation: 4% paraformaldehyde/ PB	RT	5 min
2. Two successive washes in PBS	RT	2.5 min
3. Proteinase K at 50 μg/mL in 50 mM Tris pH 8.0, 50 mM EDTA for 7.5 min	37°C	7.5 min
4. Rinse in PBS (twice)	RT	2.5 min
5. 0.05N HCl for 7.5 min	RT	7.5 min
6. Rinse in PBS	RT	2 min
7. Additional postfixation: 4% paraformaldehyde/PB	RT	5 min
8. Rinse PBS (twice)	RT	2.5 min
9. Dehydration through 60%, 80%, 95%, and 100% ethanol containing 0.33M ammonium acetate	RT	2.5 min
10. Air dry	37°C	30 min

8.4. Prehybridization

The prehybridization is to block nonspecific binding of the labeled probe. In some experiments, we have observed that omitting the prehybridization step does not result in an appreciable increase in background. However, prehybridization is recommended and should be preformed just prior to hybridization. For both the prehybridization and hybridization incubations, we have found the most convenient set up is a plexiglass box with a hinged lid approximately 10-cm deep containing one or two 3-cm wide platforms, raised about 5 cm above the base, on which the slides are placed. To maintain humidity and help prevent evaporation, wet paper towels are placed on the base. The prehybridization in this sealed container is performed without coverslips for 2–3 h at 48°C by placing it in a standard incubator. Under these conditions, there should be no apparent loss of prehybridization solution.

For both prehybridization and hybridization, a single solution is prepared from sterile stock solutions. A hybridization solution of 30 mL, sufficient for a typical experiment of 25–30 slides, is made up as follows:

	Stock	(Final)
15.0 mL	Formamide (nucleic acid grade EM Merck, BRL)	50%

	Stock	(Final)
1.8 mL	5M NaCl (autoclaved)	0.30M
0.3 mL	1M PIPES, pH 6.8 (Calbiochem, Ultrol grade, autoclaved)	0.01M
0.6 mL	0.5M EDTA, pH 7.8 (autoclaved)	0.01M
0.5 mL	3M Dithiothreitol	0.05 M
1.5 mL	100x Denhardt's solution (0.45 μm filtered) (2% BSA, 2% polyvinylpyrrolidone, 2% ficoll)	5X
0.3 mL	20%SDS (0.45 μm filtered)	0.2%
6 mL	50% Dextran Sulfate (Pharmacia)	10%
3 mL	Sterile distilled H$_2$O	

8.4.1. Preparation of Carrier RNA and DNA

The solution is filtered through a 0.45 μm Acrodisc (Gelman Sciences) disposable filter assembly. Yeast RNA and salmon sperm DNA (0.3 mL at 10 mg/ml, for a final concentration each of 100 μg/mL) are heated to 100°C for 3 min for complete denaturation and added to the hybridization solution. Both yeast RNA and salmon sperm DNA require purification before use.

Yeast RNA (Sigma type III) is dissolved at 10 mg/mL in 10 mM Tris pH 7.4 and 0.2% SDS (ETS buffer) and extracted 2–3 times with an equal vol of phenol:chloroform:isoamyl alcohol (25:24:1) saturated with 100 mM sodium acetate pH 5.2 10 mM EDTA followed by extraction with chloroform:isoamyl alcohol (24:1) and precipitation with 2 vol of ehtanol in the presence of 0.2M sodium acetate pH 5.2. The RNA is resuspended at 10 mg/mL in ETS buffer. If the absorbence ratio at 260 nm/280 nm is not between 1.7 and 2.1, the extraction is repeated.

Salmon sperm DNA (Sigma type III) must be broken down to an average single-strand length of about 250 nucleotides by either shearing, sonication, or depurination. For depurination, DNA is dissolved at 5 mg/mL in 20 mM TMS pH 7.4 by stirring overnight at 4°C. The solution is brought to 50 mM with 2M sodium acetate, pH 4.2 and heated with stirring to 70°C for 40 min. The solution is brought to 0.2N NaOH with 10N NaOH, heated further to 100°C in a boiling H$_2$O bath for 20 min, and then neutralized to pH 7.6 with HCl. The DNA is precipitated with 2 vol of ethanol, resuspended in 10 mM Tris pH 8.0, 1 mM EDTA, and extracted and reprecipitated as described for the yeast RNA.

8.4.2. Prehybridization

For prehybridization, 750 μL of the solution is applied to the slide, with care to cover each tissue section uniformly. The slides are placed in the humidified, prewarmed box and incubated at 48°C for 2–3 h. The remaining hybridization solution is stored at RT for hybridization.

8.4.3. Hybridization

The labeled RNA probe is resuspended in sterile H_2O at an estimated concentration of 2 ng/μL, typically 50 μL for a 25 μL transcription reaction. Recovery of the probe is monitored by 1 μL spotting on a nitrocellulose filter and counting in a scintillation counter. Hybridization solution is then added to a concentration of 60 ng of probe per mL (or up to 300 ng/mL for ^{35}S probes of lower specific activity) and mixed well by pipeting. The prehybridization solution is removed from the slides by carefully draining and blotting on Kimwipes and 80–100 μL of hybridization solution containing the labeled probe is distributed across the slide. A coverslip (22 × 50 mm) is carefully placed on the slide and sealed with a waterproof contact cement applied with a syringe. We have not observed any difference between siliconized or nonsiliconized coverslips and routinely use coverslips directly out of the box. We recommend Royalbond Grip Cement (Indal Aluminum Products, Los Angeles, CA) for sealing the coverslips, since it is an effective adhesive and it is easily removed with the coverslip after hybridization. The slides are returned to the incubation box for hybridization at 48°C overnight.

8.5. Posthybridization Treatments

After hybridization, the slides are immersed in 4 × SSC (1 × SSC is 0.15M NaCl, 0.015M sodium citrate and made up as a 20X stock solution) containing 300 mM β-mercaptoethanol, and the coverslips are gently pried off with pointed forceps. The high β-mercaptoethanol concentration is used to minimize nonspecific probe adherence because of disulfide bond formation, but requires that this be done in a ventilated chemical hood. The slides are then placed in a slide rack immersed in the same initial wash solution and taken through the following washes with 500 mL of the indicated solutions in staining dishes.

1. 4 × SSC without β-mercaptoethanol — RT — 15 min
2. 0.5M NaCl, 50 mM Tris pH 8.0, 5 mM EDTA containing 50 μg/mL pancreatic RNase A. RNase A is prepared at 10 mg/mL in 50 mM sodium acetate, pH 5.2, and pretreated at 80°C for 10 min — 37°C — 30 min
3. 0.5M NaCl, 50 mM Tris pH 8, 5 mM EDTA (without ribonuclease) — 37°C — 30 min
4. 2 × SSC — 56°C — 30 min
5. 0.2 × SSC — 42°C — 30 min

The residual radioactivity on the slides can be monitored, although with low efficiency, with a Mini monitor geiger counter. We are generally satisfied when the slides register 5–20 counts/s. Higher stringency washes (0.2 × SSC, 50% formamide, 50°C or 0.2 × SSC, 67°C) can be performed, but this is usually unnecessary and on occasion leads to loss of the sections. The slides are then air-dried for autoradiographic exposure.

8.6. Autoradiography

The use of ^{35}S-labeled rather than ^{3}H-labeled transcripts probes provides the opportunity to obtain an X-ray film image of the hybridization before emulsion autoradiography. In addition to providing a global image of the distribution of hybridization, the X-ray image is also used to estimate the length of time for exposure to the autoradiographic emulsion. The slides are exposed to high resolution Cronex 5 (Dupont) X-ray film in a light-sealed stainless-steel cassette at RT for 12–72 h.

For emulsion autoradiography, Kodak NTB-2 emulsion is diluted 1:1 with water. A convenient method is to distribute 25 mL aliquots of the emulsion in 50 mL graduated plastic centrifuge tubes kept at 4°C. The aliquot of emulsion is warmed to 42°C in a water bath for 45–60 min. Under a darkroom sodium safe light (Thomas Duplex Super Safelight), distilled H_2O is added to the 50-mL level and mixed well, but gently to avoid bubbles, by inversion. Let the diluted emulsion equilibrate at 42°C for an additional 30 min before dipping the slides. Initially, a blank test slide should be dipped and examined in the safe light to determine that the emulsion coating is uniform. After dipping, the slides are drained and blotted briefly on end, and set in a slide rack to dry for 1 h. The slides are then set to expose in a light-tight slide box at 4°C with dessicant. As a rule of thumb, the length of exposure to emulsion is generally four–fivefold that used to obtain an optimal signal on Cronex X-ray film.

The emulsion-coated slides are developed by standard procedures in glass staining dishes at 15–16°C using Kodak D-19 developer, diluted 1:1, for 2–3 min, a 1-min second rinse in H_2O, followed by 5 min in Kodak fixer and rinsed in running water for 10 min. The slides are then either stained with cresyl violet, or immediately dehydrated through 60%, 80%, 95%, and 100% ethanol solutions and xylene for mounting of coverslips.

Acknowledgments

We wish to thank our colleagues Ian Lipkin, Joe Watson, and especially Ellen Hess for their encouragement and reading of the manuscript during its preparation. We thank Rick Hart for his excellent technical help in doing the experiments to illustrate the optimalization of in situ hybridization shown in Fig. 3 and 4, and acknowledge G. A. Oyler and R. A. Hawkins for computer analysis of in situ hybridization shown in Fig. 6. We also especially thank Michelle Dietrich for her patient and excellent secretarial assistance in preparing the manuscript through its various revisions. This work was supported in part by grants from the National Institutes of Health NS23038 and CA33730 (M. C. W.) and a Hereditary Disease Foundation postdoctoral fellowship (G. A. H.). This is publication number 5203MB from the Research Institute of Scripps Clinic.

References

Akam M. E. (1983) The location of Ultrabithorax transcripts in Drosphila tissue sections. *EMBO J.* **2,** 2075–2084.

Amara S. G., Jonas V., Rosenfield M. G., Ong E. S., and Evans R. M. (1982) Alternative RNA processing in calculation gene expression generates mRNAs encoding different polypeptide products. *Nature (Lond.)* **298,** 240–244.

Angerer L. M. and Angerer R. C. (1981) Detection of poly A$^+$ RNA in sea urchin eggs and embryos by quantitative *in situ* hybridization. *Nucl. Acids Res.* **9,** 2819–2840.

Angerer L. M., Stoler M. H., and Angerer R. C. (1987) *In situ* hybridization with RNA probes: An annotated recipe, in *In Situ Hybridizations: Applications to Neurobiology* (Valentino K. L., Eberwine J. H., and Barchas J. D., eds.), Oxford University Press, New York, pp. 42–70.

Arquint M., Roder J., Chia J., Down D., Wilkinson D., Bayley H., Braun P., and Dunn J. (1987) Molecular cloning and primary structure of myelin-associated glycoprotein. *Proc. Natl. Acad. Sci. USA* **84(2),** 600–604.

Bonner T. I., Brenner D. J., Neufeld B. R., and Britten R. J. (1973) Reduction in the rat of DNA reassociation by sequence divergence. *J. Mol. Biol.* **81,** 123–135.

Brahic M. and Haase A. T. (1978) Detection of viral sequences of low reiteration frequency by *in situ* hybridization. *Proc. Natl. Acad. Sci. USA* **75,** 6125–6129.

Brahic M., Haase A. T., and Cash E. (1984) Simultaneous *in situ* detection of viral RNA and antigens. *Proc. Natl. Acad. Sci. USA* **81,** 5445–5448.

Branks P. and Wilson M. C. (1986) Patterns of gene expression in the murine brain revealed by *in situ* hybridization of brain specific messenger RNAs. *Mol. Brain Res.* **1,** 1–19.

Brigati D. J., Myerson D., Leary J. J., Spalholz B., Travis S. Z., Fong C. K. Y., Hsiung G. D., and Ward D. C. (1983) Detection of viral genomes in cultured cells and paraffin-embodded tissue sections using biotin-labeled hybridization probes. *Virology* **126,** 32–50.

Buongiorno-Nardelli M. and Amaldi F. (1970) Autoradiographic detection of molecular hybrids between rRNA and DNA in tissue sections. *Nature (Lond.)* **225,** 946–948.

Casey J. and Davidson N. (1977) Rates of formation and thermal stabilities of RNA:DNA and DNA:DNA duplexes at high concentrations of formamide. *Nuc. Acids Res.* **4,** 1539–1552.

Chamberlin M. and Ryan T. (1982) Bacteriophage DNA-dependent RNA Polymerases, in *The Enzymes* (Boyer P. D., ed.), Vol. XV, Part B, Academic Press, New York, pp. 87–108.

Coghlan J. P., Aldred P., Naralambidis J., Niall H. D., Penschow J. D., and Tregear G. W. (1985) Hybridization histochemistry. *Anal. Biochem.* **149,** 1–28.

Cox K. H., deLeon D. V., Angerer L. M., and Angerer R. C. (1984) Detection of mRNAs in sea urchin embryos by *in situ* hybridization using asymmetric RNA probes. *Dev. Biol.* **101,** 485–502.

Darnell J. E. and Wilson M. C. (1982) Regulatory mechanisms of gene expression in higher eukaryotes, in *Molecular Genetic Neuroscience* (Schmitt, F. O., Bird, S. J., and Bloom, F. E., eds.), Raven, New York, pp. 25–35.

Dunn A. R., Gallimore P. H., Jones K. W., and McDougall J. K. (1973) *In Situ* hybridization of adenovirus RNA and DNA II. Detection of adenovirus-specific DNA in transformed and tumour cells. *Int. J. Cancer* **11,** 628–636.

Ehn E. and Larsson B. (1979) Considerations of the use of Ultrofilm *Science Tools* **26,** 24–29.

Feinberg A. P. and Vogelstein B. (1983) A technique for radiolabeling DNA restriction fragments to high specific activity. *Anal. Biochem.* **132,** 6–13.

Fink J. S., Montiminy M. R., Tsukada T., Hoefler H., Specht L., Lechan R. M., Wolfe H., Mandel G., and Goodman R. H. (1986) In situ hybridization of somatostatin and VIP mRNA in the rat nervous system, in *In Situ Hybridization in Brain* (Uhl G. R., ed.), Plenum, New York. pp. 181–191.

Flavell R. H., Birfelder E. J., Sanders J. P. M., and Borst P. (1974) DNA-DNA hybridization on nitedce Nulose filters: 1. General considerations and non-ideal kinetics.

Fremeau R. T., Lundblad J. R., Pritchett D. B., Wilcox J. N., and Roberts J. L. (1986) Regulation of pro-opiomelanocortin gene transcription in individual cell nuclei. *Science* **234,** 1265–1269.

Gall J. and Pardue M. (1969) Formation and detection of RNA-DNA hybrid molecules in cytological preparations. *Proc. Natl. Acad. Sci. USA* **63,** 378–383.

Gee C. E. and Roberts J. L. (1983) *In situ* hybridization histochemistry: A technique for the study of gene expression in single cells. DNA **2,** 157–163.

Gee C. E., Chen C. -L. C., Roberts J. L., Thompson R., and Watson S. J. (1983) Identification of proopiomelanocortin neurones in rat hypothalamus by *in situ* cDNA-mRNA hybridization. *Nature (Lond.)* **306,** 374–376.

Godard C. M. and Jones K. W. (1979) Detection of AKR MuLV-specific RNA in AKR mouse cells by *in situ* hybridization. *Nuc. Acids Res.* **6,** 2849–2861.

Godard C. M. and Jones K. W. (1980) Improved method for detection of cellular transcripts by in situ hybridization: Detection of poly (A) sequences in individual cells. *Histochem.* **65,** 291–300.

Goldman D., Simmons D., Swanson L., Patrick J., and Heineman S. (1986) Mapping brain areas expressing RNA homologous to two

different acetylcholine receptor alpha subunit cDNAs. *Proc. Natl. Acad. Sci. USA* **83,** 4076–4080.
Green M., Maniatis T., and Melton D. (1983) Human beta globin pre-mRNA synthesized *in vitro* is accurately spliced in xenopus oocyte nuclei. *Cell* **32,** 681–694.
Haase A. (1987) Analysis of viral infections by *in situ* hybridization, in *In Situ Hybridization: Applications to Neurobiology* (Valentino K. L., Eberwine J. H., and Barchas J. D., eds.), Oxford University Press, New York, pp. 197–220.
Haase A. T., Stowring J. D., Harris B., Traynor B., Ventura P. M., Peluso R., and Brahic M. (1982) Visna DNA synthesis and the tempo of infection *in vitro, Virology* **119,** 399–410.
Haase A. T., Walker D., Stowring L., Ventura P., Geballe A., and Blum H. (1985) Detection of two viral genomes in single cells by double-label hybridization *in situ* and color microradioautography. *Science* **227,** 189–191.
Hafen E., Levine M. L., and Garber R. L. (1983) An improved *in situ* hybridization method for the detection of cellular RNAs in Drosophila tissue sections and its application for localizing gene transcripts of the homeotic Antennapedia gene complex. *EMBO J.* **2,** 617–623.
Harrison P. R., Conkie D., Paul J., and Jones K. (1973) Localization of cellular globin messenger RNA by *in situ* hybridization to complimentary DNA. *FEBS Letts.* **32,** 109–112.
Hibbard L. S., McGlone J. S., Davis D. W., and Hawkins R. A. (1987) Three-dimensional representation and analysis of brain energy metabolism. *Science* **236,** 1641–1646.
Higgins G. A., Lewis D. A., Bahmanyar S., Goldgaber D., Gajdusek D. C., Young W. G., Morrison J. H., and Wilson M. C. (1988) Differential regulation of amyloid-β-protein mRNA expression within hippocampal neuronal subpopulations in Alzheimer's disease. *Proc. Natl. Acad. Sci. USA* **85,** 1297–1301.
Hudson P., Penschow J. D., Shine J., Ryan G., Niall H. D., and Coghlan J. P. (1981) Hybridization histochemistry: Use of recombinant DNA as a "homing probe" for tissue localization of specific mRNA population. *Endocrin.* **108,** 353–356.
John H. A., Birnstiel M. L., and Jones K. W. (1969) RNA-DNA hybrids at the cytological level. *Nature (Lond.)* **223,** 582–587.
Kuhar M. J. and Unnerstall J. R. (1985) Quantitative receptor mapping by autoradiography: Some current technical problems. *Trends NeuroSci.* **8,** 49–53.
Lai C., Brow M. A., Nave A. B., Noronha A. B., Quarles R. H., Bloom F. E., Milner R. J., and Sutcliffe J. G. (1987) Two forms of 1B236/myelin-

associated glycoprotein, a cell adhesion molecule for postnatal neural development, are produced by alternative splicing. *Proc. Natl. Acad. Sci. USA* **84(12)**, 4337–4441.

Langer P R., Waldrop A. A., and Ward D. C. (1981) Enzymatic synthesis of biotin-labeled polynucleotides: novel nucleic acid affinity probes. *Proc. Natl. Acad. Sci. USA* **78**, 6633–6637.

Lawrence J. B. and Singer R. H. (1986) Intracellular localization of mRNAs for cytoskeletal proteins. *Cell* **45**, 407–415.

Lawrence J. B. and Singer R. H. (1985) Quantitative analysis of *in situ* hybridization methods for the detection of actin gene expression. *Nuc. Acids Res.* **13**, 1777–1799.

Lawrence J. B., Villnave C. A., and Singer R. H. (1988) Sensitive, high resolution chromatin and chromosome mapping *in situ:* Presence and orientation of two closely integrated copies of EBV in a lymphoma line. *Cell* **52**, 51–61.

Levinthal C. and Ware R. (1972) Three-dimensional reconstruction from serial sections. *Nature (Lond.)* **236**, 207–210.

Lewis M. E., Sherman T. G., and Watson S. J. (1985) *In Situ* hybridization histochemistry with synthetic oligonucleotides: Strategies and methods. *Peptides* **6**, **(Suppl. 2)**, 75–89.

Lewis M. E., Sherman T. G., Burke S., Akil H., Davis L. G., Arentzen R., and Watson S. J. (1986) Detection of proopiomelanocortin mRNA by *in situ* hybridization with an oligodeoxynucleotide probe. *Proc. Natl. Acad. Sci. USA* **83**, 5419–5423.

Lynn D. A., Angerer L. M., Brushkin A. M., Klein W. H., and Angerer R. C. (1983) Localization of a family of mRNAs in a single cell type and its precursors in sea urchin embryos. *Proc. Natl. Acad. Sci. USA* **80**, 2656–2660.

Maniatis T., Fritsch E. F., and Sambrook J. (eds.) (1982) *Molecular Cloning: A Laboratory Manual.* Cold Spring Harbor Laboratory, New York.

McConaughy B. L., Laird C. D., and McCarthy B. J. (1969) Nucleic acid reassociation in formamide. *Biochem.* **8**, 3289–3295.

Miller F. D., Naus C. C. G., Higgins G. A., Bloom F. E., and Milner R. J. (1987) Developmentally regulated rat brain mRNAs: Molecular and anatomical characterization. *J. Neurosci.* **7**, 2433–2444.

Nobou F., Benson J., Rostene W., and Rosselin G. (1985) Ontogeny of vasoactive intestinal peptide and somatostatin in different structures of rat brain: effects of hypo- and hypercorticism *Dev. Brain Res.* **20**, 296–301.

Orth G., Jeanteur P., and Croissant O. (1970) Evidence for and localization of vegatative viral DNA replication by autoradiographic detection of RNA–DNA hybrids in sections of tumors induced by shape papilloma virus. *Proc. Natl. Acad. Sci. USA* **68**, 1876–1880.

Quarles R. H. (1984) Myelin-associated glycoprotein in development and disease. *Dev. Neurosci.* **6**, 285–303.

Rigby P. W. J., Dieckmann M., Rhodes C., and Berg P. (1977) Labeling deoxyribonucleic acid to high specific activity *in vitro* by nick-translation with DNA polymerase I. *J. Mol. Biol.* **113**, 237–251.

Rogers A. W. (1979) *Techniques of Autoradiography*. 3rd Ed. Elsevier North Holland, Amsterdam.

Schmale H. and Richter D. (1984) Single base deletion in the vasopressin gene is the cause of diabetes insipidus in Brattleboro rats. *Nature (Lond.)* **308**, 705–709.

Shivers B. D., Harlan R. E., Pfaff D. W., and Schacter B. S. (1986) Co-localization of peptide hormones and peptide hormone mRNAs in the same tissue sections of rat pituitary. *J. Histochem. Cytochem.* **34**, 39–43.

Singer R. H. and Ward D. C. (1982) Actin gene expression visualized in chicken muscle tissue culture by using *in situ* hybridization with a biotinated nucleotide analogue. *Proc. Natl. Acad. Sci. USA* **79**, 7331–7335.

Szabo P., Elder R., Steffensen D. M., and Uhlenbeck O. C. (1977) Quantitative *in situ* hybridization of ribosomal RNA species to polytene chromosomes of Drosophila melanogaster. *J. Mol. Biol.* **115**, 539–563.

Thomas C. A., Jr. and Dancis B. M. (1973) Ring stability. Appendix to: Formation of rings from Drosophila DNA fragments. *J. Mol. Biol.* **77**, 44–55.

Uhl G. R. (ed.) (1986) *In Situ Hybridization in Brain*. Plenum, New York.

Uhl G. R., Zingg H. H., and Habener J. F. (1985) Vasopressin mRNA *in situ* hybridization: Localization and regulation studied with oligonucleotide cDNA probes in normal and Brattleboro rat hypothalamus. *Proc. Natl. Acad. Sci. USA* **82**, 5555–5559.

Utset M. F., Awgulewitsch A., Ruddle F. H., and McGinnis W. (1987) Region-specific expression of 2 mouse homeobox genes, *Science* **235**, 1379–1382.

Valentino K. L., Eberwine J. H., and Barchas J. D. (eds.) (1987) *In Situ Hybridization: Applications to Neurobiology*. Oxford University Press, New York.

Wahl G. M., Stern M., and Stark G. R. (1979) Efficient transfer of large DNA fragments from agarose gels to diazobenzyloxymethyl-paper and rapid hybridization by using dextran sulfate. *Proc. Natl. Acad. Sci. USA* **76**, 3683–3687.

Wallace R. B. and Miyada C. G. (1987) Oligonucleotide Probes For Screening Of Recombinant DNA Libraries, in *Methods in Enzymology* **152**. (Berger S. C. and Kimmel A. R., eds.), Academic Press, New York, pp. 433–442.

Wallace R. B., Johnson M. J., Hirose T., Miyake T., Kawashima E. H., and Itakura K. (1981) The use of synthetic oligonucleotides as hybridization probes. *Nuc. Acids Res.* **9**, 879–894.

Wetmur J. G. and Davidson N. (1968) Kinetics of renaturation of DNA. *J. Mol. Biol.* **31**, 349–370.

Wetmur J. G., Ruyechan W. T., and Douthart R. J. (1981) Denaturation and renaturation of Penicillium chrysogenum mycophage double-stranded ribonucleic acid in tetra-alkylammonium salt solutions. *Biochem.* **20**, 2999–3002.

Wolfson B., Manning R. W., Davis L. G., Arentzen R., and Baldino F. (1985) Co-localization of corticotropin releasing factor and vasopressin mRNA in neurons after adrenalectomy. *Nature (Lond.)* **315**, 59–61.

Wood W. I., Gitschier J., Lasky L. A., and Lawn R. M. (1985) Base composition-independent hybridization in tetramethylammonium chloride: A method for oligonucleotide screening of highly complex gene libraries. *Proc. Natl. Acad. Sci. USA* **82**, 1585–1588.

Young W. S. (1986) Quantitative *in situ* hybridization and determination of mRNA content, in *In Situ Hybridization in Brain* (Uhl G., ed.), Plenum, New York, pp. 243–248.

Young W. G., Morrison J. H., and Bloom F. E. (1985) An electronic morphometry and mapping analysis microscopy system (EMMA) for the quantitative and comparative study of neural structures. *Soc. Neurosci. Abstr.* **11**, 670.

zur Hausen H. and Schultz-Holthausen H. (1972) Detection of Epstein-Barr Viral genomes in human tumour cells by nucleic acid hybridization, in *Oncogenesis and Herpes* (Biggs P. M., de-The G., and Payne L. N., eds.), I.A.R.C., Lyons, France, pp. 321–325.

Index

Abelson-transformed B cells, 205
Acetic anhydride, 263
ACTH, 240
Actin, 259
Actinomycin D, 16, 26
Acrylamide, 42, 107, 126, 127, 144, 146
S-Adenosyl methionine, 29
β-Adrenergic receptors, 89
Agarose, 11, 28, 29, 65, 122, 126, 138
Ala (see Alanine)
Alanine (Ala), 4, 87, 91, 99
Alcohol dehydrogenase, 196
ALIGN, 95
Alignment score, 95
Alkaline phosphatase, 22, 36, 143
[Alpha-^{32}P]dCTP, 25, 26
[Alpha-^{35}S]dATP, 38
ALV (see Avian leukemia virus)
Alzheimer's disease, 71, 74, 268
γ-Aminobutyric acid, (see GABA)
Ammonium persulfate, (APS) 42
Ampicillin, 20, 65, 66
Ampulla, 222, 223
AMV (see Avian myeloblastosis virus)
AMV reverse transcriptase, 25, 26, 67
Amyloid-β-protein, 268
Antibiotin-IgG, 259
Antisense or minus strand, 84
Antisense RNA, 232
APS (see Ammonium persulfate)
Arg (see Arginine)
Arginine (Arg), 4, 91, 92, 98
Asialoglycoprotein receptor, 89
Asn (see Asparagine)

Asp (see Aspartic acid)
Asparagine (Asn), 4, 90
Aspartic acid (Asp), 4, 98
ATP, 5, 32, 34, 41, 123, 130, 254, 258, 272
Autographa californica nuclear polyhedrosis virus, 105
Autoradiography, 29, 33, 41, 58, 60, 69, 70, 72, 73, 81, 140, 146, 161, 263–266, 277
Avian leukemia virus (ALV), 182, 202
Avian myeloblastosis virus (AMV), 14, 144
2,2'-Azino-di[3-ethylbenzthiazolin-sulfonate(6)]diammonium salt, 102

1B236, 243
1B236/MAG, 87, 88, 90, 96, 106
B82 (GM 0347A), 158
Baclovirus, 105
Bacterial expression systems, 103
Bacteriophages (see also Phages), 19, 37, 49, 57, 82, 271
Bacteriophage SP6, 19
Bactrotryptone, 38
BAG virus, 181, 194, 195, 201, 202, 204
Bam HI site, 121, 138, 140, 141
BioRex resin, 42
Biotin, 259
Biotin-derivatized nucleotides, 82, 258
Biotinylated anti-IgG, 262
Biotinylated probes, 258

285

Bis-acrylamide, 42
Bisacrylocysteine, 127
Blunt-end ligation, 17
Bovine serum albumin (BSA), 28–30, 101, 102, 131, 133, 249, 271
Brattleboro rat, 262
5-Bromo-4-chloro-3-indolyl beta D-galactoside (X-gal), 35, 38, 196–200
Bromophenol blue, 40, 127, 128
BSA (*see* Bovine serum albumin)

CAAT element, 146, 147
Calcitonin, 118
Calcitonin gene-related peptide (CGRP), 118
Calcium, 90, 91, 96, 104, 147
Calmodulin, 56, 57, 90, 91, 96
cAMP, 90, 91
Candidate gene hypothesis, 170
Carboxyfluorescein diacetate succinimyl ester, 200
Carnoy's fixative, 250
Carrier RNA, 275
Casamino acids, 10
CAT (*see* Chloramphenicol acetyltransferase)
Cataracts, 170
cDNA (*see* Complementary DNA)
CE201, 125
Centimorgans, 156
Cerebellum, 65, 70–75, 194, 199, 208, 243, 251, 259
Cerebral cortex, 65–67, 70–76, 194, 200, 208, 251, 257
CGRP (*see* Calcitonin gene-related peptide)
Charon vectors, 121
Chloral hydrate, 269
4-Chloro-1-naphthol, 108

Chloramphenicol acetyltransferase (CAT), 147
Choroid plexus, 234
Chromatids, 259
Chromogranin B, 94
Chromosome 4, 172
Chromosome 7, 172
Chronic myelogenous leukemia, 154
Cis-regulatory elements, 230
Clones, (*see also* Complementary DNA), 13–43, 49–76, 79, 80, 82, 84, 85, 103, 105, 108, 110, 119, 121, 124–126, 131–136, 138–147, 159, 160, 163, 202–204, 221, 232
Coding region sequences, 92
Codon, 85
Collagen genes, 171
Colony hybridization, 21
Complementary DNA (cDNA), 1, 13–20, 22–24, 26–29, 33, 34, 37–39, 50–61, 63–66, 68, 71–74, 79–82, 84, 85, 97, 98, 103, 104, 106, 125, 141, 145, 160, 180, 192, 194, 241, 252, 253, 255, 258, 261, 271
  preparation and use of subtractive cDNA hybridization probes for cDNA cloning, 49–76
  preparation of cDNA libraries and isolation and analysis of specific clones, 13–43
Complementary RNA (cRNA), 54, 57, 82, 241
Consensus sequences, 85, 86
Contig mapping, 156
Corpus callosum, 245
COS 7, 104
Cosmids, 19, 121
CRE, 181, 182, 190
Creatine phosphate, 5, 6

Creatine phosphokinase, 5
m-Cresol, 120
Cresyl violet, 268, 278
CRIP, 181, 182, 190, 207
CRIP-myc, 189
cRNA (see Complementary RNA)
Cryomicrotomy, 270
γ Crystalline gene, 170
CsCl/ethidium bromide equilibrium density gradient, 270, 271
C-test, 39
CTP, 254, 257, 272
Cyclic AMP (see cAMP)
Cys (see Cysteine)
Cysteine (Cys), 4, 5, 93
Cystic fibrosis, 172

D17, 190
D17–C3, 181, 182, 190
DAB (see Diaminobenzidine tetrahydrochloride)
dATP, 41, 67, 131
dCTP, 39, 40, 67, 71, 131, 145
ddA, 41
ddC, 41
ddG, 41
ddT, 41
DEAE, 147
DEAE-cellulose, 126, 129
DEAE-dextran, 184
DEAE-nylon, 126
Denhardt's solution, 133, 160, 161, 249, 275
Densometric analysis, 264
Dentate gyrus (DG), 243, 245, 257, 259
Deoxynucleotides (dNTPs), 14, 25, 26, 28, 30, 39
Deoxynucleotide transferase, 75
Deoxyribonucleic acid (see DNA)

Deoxytrinucleotide mixture (DTM), 131
Dextran sulfate, 161, 246, 247, 275
DG (see Dentate gyrus)
dGTP, 39, 40, 67, 131, 145
Diabetes, 171
DIAGON, 95
Diaminobenzidine tetrahydrochloride (DAB), 262, 263
Dideoxynucleotide stocks (ddNTPs), 40
Diethylpyrocarbonate, 273
Differential (plus-minus) colony hybridization, 58
Digitonin, 250
Dimethylformamide, 197
Disulfide-acrylamide gels, 126–128
Dithiothreitol (DTT), 3–5, 7, 8, 11, 25, 26, 30–32, 34, 41, 67, 123, 131, 255, 272, 273
DMD (see Duchenne's muscular dystrophy)
DME, 185, 187, 188, 191, 199
DNA (see also Complementary DNA), 15, 178, 179, 181, 183, 184, 192, 194, 221, 223–229, 233, 239, 245, 246, 252, 253, 255, 270, 271, 273, 275
  kinase, 31
  ligase, 16, 17, 28, 31, 32, 34, 123
  polymerase, 27, 28, 30, 41, 130, 131, 252, 253, 258
DNase, 8, 130, 131, 137, 252, 257, 273
dNTPs (see Deoxynucleotides)
DO-L, 193
Dorsal root ganglion, 194
Dot matrix analysis, 95
Drosophila, 97, 164, 165, 196

ds-cDNA, (see also Complementary DNA) 15–17, 29, 30, 32, 34
ds-DNA, 17, 247, 252–254, 258, 260, 268
DTM (see Deoxytrinucleotide mixture)
DTT (see Dithiothreitol)
dTTP, 39, 40, 131, 145
Duchenne's muscular dystrophy (DMD), 118, 119, 171
Dystrophin, 119

E10-E14 mouse embryos, 201
E12 embryos, 200
E13 embryos, 200
E. coli, 252, 253, 258
 LE 392, 123, 125
 Y1090, 35, 36
 Y1090 (hsdR-), 35
EcoRI, 17, 18, 29, 32, 34, 38, 39, 66, 122, 138, 140, 141, 159
 linkers, 29, 32
 mediated ligation, 17
 methylase, 29
Edestin, 99, 100
E–F hand, 96
Elastase gene, 231
Electroelution, 126, 130
Electrophoresis, 11, 27–29, 32, 39, 41, 42, 51, 65, 69, 107, 122, 126, 127, 138, 144, 146
ELISA, 101, 102
EMBL vectors, 120–122, 125
EMMA system, 109, 267
Endoplasmic reticulum, 87, 90
Entopeduncular nucleus (EP), 243
Env genes, 181, 182
EP (see Entopeduncular nucleus)
Epidermal growth factor, 97
Epitopes, 56, 98

Ethanol, 275
Ethidium bromide, 138
Eukaryotic expression systems, 104
Exons, 117, 118, 125, 145

Families, 96
FASTP, 93
Fatty acids, 92
Fibroblasts, 157, 158, 180, 202, 205, 209
Ficoll, 249
Fluorescein-avidin, 259
Fluorescein di-β-D-galactopyranoside, 204
Fluorescent microscope, 200
Formamide, 40, 41, 146, 160, 161, 245, 255, 274
Formamide-dye mix, 40, 41
Fornix, 245
Fourth complement component, 57
Freund's adjuvant, 100

G418, 187–190, 192, 208–210
GABA, 243
GAD (see Glutamate decarboxylase)
Gag genes, 179, 180
β-Galactosidase (β-gal), 19, 22, 35, 38, 56, 104, 195, 199, 204
Gemini vectors, 82
Genes (see also Clones)
 genetic linkage analyses, 163–173
 in situ hybridization, 239–278
 isolation and structure determination of, 117–147
 lineage analysis and immortalization of neural cells, 177–211

Index

mapping techniques, 153–163
transgenic mice, 221–234
Genomic libraries, 119, 120
Glioma cell lines, 194
Gln (*see* Glutamine)
Globus pallidus, 243
Glu (*see* Glutamic acid)
Glucocorticoids, 117, 241
Glucose, 266
Glutamate (Glu), 4, 56, 98, 211, 243
Glutamate decarboxylase (GAD), 56, 243
Glutamine (Gln), 4, 91
Glutaraldehyde, 100, 196, 197, 249, 263
Gly (*see* Glycine)
Glycerol, 127, 143, 185
Glycine (Gly), 4, 87, 91, 92, 99
Glycoprotein, 181–183
Glycosylation, 90, 99, 103, 105, 182
GnRH (*see* Gonadotrophin-releasing hormone)
Golgi apparatus, 87, 90
Gonadotrophin-releasing hormone (GnRH), 234
GP+E-86, 181, 190
Group A proteins, 88
Group B proteins, 89
Group C proteins, 89, 90
Growth factors, 208
Growth hormone, 118, 240, 261
GTP, 5, 6, 254, 272

Hae III buffer, 138
Hardy-Weinberg frequencies, 169
HAT medium, 159
HBS, 184, 185
hCG (*see* Human chorionic gonadotropin)

HD (*see* Huntington's disease)
Hematoxylin/eosin, 268
HEPES, 3, 5, 6, 185, 273
Herpes simplex virus (HSV), 205, 206
Heterochromatin, 154
Hind III, 140, 159
Hippocampus, 245, 251, 257, 259, 268
His (*see* Histidine)
Histidine (His), 5, 98
HnRNA transcripts, 241
Hoffman modulation contrast optics, 224
Homology, 93, 94
Horseradish peroxidase, 22, 108, 262
Howley's buffer, 138
HPRT, 157
HRAS, 172
HSV (*see* Herpes simplex virus)
Human chorionic gonadotropin (hCG), 222
Huntington's disease (HD), 118, 165, 169, 171
Hybrid-selected translation, 22, 23
Hydrogen peroxide, 108
Hydroxyapatite chromatography, 60, 62, 64, 67, 69, 75
8-Hydroxyquinoline, 120
Hyluronidase, 223
Hypothalamus, 240
Hypoxanthine-guanine phosphoribosyl transferase (HPRT), 229

Ile (*see* Isoleucine)
Immortalization of neural cells, 204–211

Immunoaffinity chromatography, 97, 109
Immunocytochemistry, 262, 263, 268
Immunoglobulin, 104
Immunoglobulin gene superfamily, 96
IMR 91 strain, 158
Inferior colliculus, 243
In situ hybridization, 49, 65, 80, 82–84, 160, 239–278
Insulin, 172, 232
β-Interferon, 105
Internal capsule, 243
IPTG (*see* Isopropyl-beta-D-thiogalactopyranoside)
Isoleucine (Ile), 5, 87
Isopropyl-beta-D-thiogalactopyranoside (IPTG), 35, 36, 38
Ix TBE, 43

JC virus, 234
JM103, 38

Ketamine, 269
Keyhole limpet haemocyanin (KLH), 99, 100
Kinased linkers, 32
Kinasing, 130
Klenow fragment, 30, 41, 252
KLH (*see* Keyhole limpet haemocyanin)

Lac operon, 35
LacZ, 19, 22, 35, 36, 56, 195, 197, 202, 205, 211
Lambda DNA, 29
Lambda gt11, 19, 20, 24, 25, 29, 34, 38, 56, 57, 104

Lambda J1, 121
Lambda ZAP, 82
LB broth, 123, 143
Leu (*see* Leucine)
Leucine (Leu), 5, 87
Lin-12, 97
Lineage mapping, 194–204
LINKAGE, 168
Linkage analysis, 153, 154, 163–174
LIPED, 168
Locus ceruleus, 83
Lon protease, 36
Long terminal repeats (LTRs), 178, 180, 183, 192–194, 202, 207
LS buffer, 131
LTRs (*see* Long terminal repeats)
Luciferase, 196
Lymphocytes, 157, 159
Lys (*see* Lysine)
Lysine (Lys), 5, 91, 98, 249

M13, 24, 37, 39, 54, 85, 144, 146, 252
M13mp19, 38, 39
Manic-depressive psychosis, 171
Mbo I, 121, 122
Meiosis, 163, 222
Melanoblast, 205
β-Mercaptoethanol, 107, 120, 126, 129, 131, 138, 255, 276, 277
Mesencephalon, 208
Messenger RNA (mRNA), 1, 8, 9, 13–15, 20, 22, 23, 49–61, 63, 66, 70, 72–76, 79–86, 90, 103, 105, 110, 117, 118, 145, 179, 239–241, 243, 245, 254, 258–265, 268, 271
Met (*see* Methionine)
Methionine (Met), 5, 9, 55

Index

7-Methylguanosine, 85, 86
Microdensitometer, 265
Microsomes, 2, 6
Mitochondria, 1
Mitomycin, 189
Mitosis, 210
MMLV (see Mouse Moloney leukemia virus)
Monoclonal antibodies, 99
Mouse Moloney leukemia virus (MMLV), 14, 181, 182, 205
Mousetree, 228
mRNA (see Messenger RNA)
MuBr8, 265, 266
MuBr8mRNA, 257
Muscarinic receptors, 89
Mutagenesis procedures, 49
Myelin, 89, 94, 96, 234, 243, 245
Myelin-associated glycoprotein (MAG), 245
Myelin basic protein (MBP), 234

NBRF-PIR, 92
N-CAM, 88, 96
Neural crest, 194
Neuroblastoma vector, 194
Nick translation, 130, 159, 252, 261, 268
NIH3T3 cells, 187, 190, 191
Nitrocellulose, 21, 80, 107, 132, 135, 139, 140, 160, 276
Nomarski differential interference contrast optics, 224
Northern blot, 23, 49, 55, 59, 71–75, 79–84, 186
Notch, 97
NP-40, 108
Nuclear RNA, 50
Nuclease, 15, 62

Nucleotide sequence complexity, 51

OCT compound, 198
Olfactory bulb, 194, 199, 208, 251
Oligodendrocytes, 243, 250
Oligo(dT), 53, 67
Oligo(dT)cellulose, 50, 65
Oligonucleotide hybrids, 258
Oncogene, 232
Oocytes, 57, 105, 222
Open reading frame (ORF), 84, 85, 87, 97, 106
Opsin, 89, 90
ORF (see Open reading frame)
Osteogenesis imperfecta, 171, 233
Ouabain, 159

PA12, 181, 190
PA317, 181, 182, 186
Packaging lines, 181–183, 190, 202
Palindromic sequences, 124
PAM250, 93, 95
Pancreas, 232
Paraformaldehyde, 197, 198, 249, 260, 261, 263, 268, 269, 284
PB (see Phosphate buffer)
pBluescript, 141, 144
pBR322, 19–21, 141
PBS (see Phosphate buffered saline)
pBS vectors, 82
PCR (see Polymerase chain reaction)
Pentobarbital, 269
Pepstatin, 107
Peroxidase, 36
PERT (phenol emission reassociation technique), 66–68, 74, 75
PERT hybridization, 64

pGEM, 141, 144, 253
pGEM4 vector, 65
Phages (see also Bacteriophages), 20, 35, 54, 57–59, 124–126, 132, 136, 137, 197
Phe (see Phenylalanine)
Phenol, 39, 62, 63, 68, 120, 129, 130, 159, 224, 227
Phenylalanine (Phe), 5, 94, 96
Phenylalanine hydroxylase, 94, 96
Phenylhydrazine, 2
Phenylmethylsulfonyl fluoride (PMSF), 7
Phosphate buffer (PB), 269, 274
Phosphate buffered saline (PBS), 100–102, 196, 198, 199, 250, 274
Phosphokinase, 6
PIPES, 146, 275
Pituitary gland, 240, 260
PL, 104
Plaque, 59
Plaque hybridization, 21
Plasmids, 18, 20, 49, 54, 58, 59, 65–67, 72, 82, 121, 125, 126, 135, 141, 142, 180, 253, 257, 270–272
Pluripotent embryonic stem (ES) cells, 229
PMC1H8a, 71, 74
PMC2A1, 72
PMC2C8, 72, 73
PMC2D1, 72
pMC2D6, 73
pMC5B3, 73
pMC4G8, 73
PMS (see Pregnant mare's serum)
PMSF (see Phenylmethylsulfonyl fluoride)
P_o, 96
Pol genes, 180

Poly (A)$^+$, 15, 82
Polyacrylamide gel, 32, 56
Poly (A)$^-$mRNA, 50
Poly (A)$^+$RNA, 25, 50, 51, 61–65, 67, 71, 73, 146
Polybrene, 184, 187, 191, 208
Polyclonal antibodies, 99
Polyethylene glycol, 158, 183
Polyhedrin protein, 105
Polylinkers, 19
Poly-L-lysine, 249
Poly-L-ornithine, 208
Polymerase chain reaction (PCR), 201, 203
Polymerase I, 15
Polymerase reaction buffer (PRB), 39, 40
Polyvinylpyrrolidone, 133, 249, 275
POMC (see Proopiomelanocortin)
Posthybridization, 276
PRB (see Polymerase reaction buffer)
Pregnant mare's serum (PMS), 222
Prehybridization, 276
Preprosomatostatin, 75
Primer extension, 145
Pro (see Proline)
Producer cells, 186, 189, 208
Prolactin, 118
Proline (Pro), 5, 91, 92
Promega, 82
Promoter elements, 146
Promoters, 193
Proopiomelanocortin (POMC), 240, 241
Protease inhibitors, 107
Protein A, 104
Proteinase K, 120, 137, 159, 227, 250, 268, 274
Protein kinases, 90, 91

Protein translation systems, 109
Proteolytic cleavage sites, 91
pSC101, 18
PseqIP, 92
Pseudogenes, 154, 160
Pseudopregnant female, 225, 226
Pseudopregnant mouse, 230
ψ, 183, 190
ψ2, 182, 186, 189, 190, 207
ψ2-BAG, 189, 199
ψam, 181, 182, 186, 190
ψ-minus, 192
ψ Packaging sequence, 180
PstI site, 20, 66
pUC, 142
pUC18/19, 141
pUC118/119, 141, 144
pUC plasmids, 20, 23, 24
Purkinje cells (PC), 243
Pyroglutamate, 92

Q2bn, 182, 202

Radioimmunoassay, 97, 109
Random oligonucleotide primers, 268
Random prime labeling, 131
Random priming, 130, 159
RecA$^+$, 125
RER (*see* Rough endoplasmic reticulum)
Restriction endonuclease, 16, 17, 20, 159, 160, 253, 271
Restriction enzymes, 22, 146, 224, 227
Reticular thalamic nucleus, 243
Reticulocyte lysate, 2
Retina, 194, 202, 203
Retroviruses, 177, 179, 180, 194, 196, 202, 204–206, 230
Reverse transcriptase, 15, 22, 50, 55, 144, 190

RFLPs, 168–170
Ribonuclease, 268
Riboprobes, 82
Ribosomal RNA, 50, 86
Ribosomes, 86, 104
RM (*see* Rough microsomes)
RNA (*see also* Complementary RNA, Messenger RNA, Nuclear RNA, Ribosomal RNA, Single-stranded RNA, and Transfer RNA), 14, 15, 27, 177, 192, 199, 230, 245–250, 254, 255, 260, 261, 263, 264, 272, 273, 275
 polymerase, 22, 57, 82, 243, 253, 271, 272
RNase, 6, 8, 120, 137, 159, 161, 255, 260, 261, 270, 273, 277
RNase H, 14, 15, 28
RNase inhibitor, 67
RNase protection, 49
RNasin, 25, 26
Rough endoplasmic reticulum (RER), 1
Rough microsomes (RM), 1
Rous Sarcoma Virus, 202

S phase, 179
S1 nuclease, 30, 145, 146
Sal I, 121, 122, 138
Scrapie, 57
SDS (*see* Sodium dodecyl sulfate)
SEARCH, 93
Sense or plus strand, 84
Sephadex, 3, 4, 100, 131, 273
Sepharose, 11, 33
Ser (*see* Serine)
Serine (Ser), 5, 87, 90, 99
Shiverer mice, 226, 234
Shotgun cloning, 141, 142
SI nuclease, 15
Signal peptide sequences, 87, 89

Signal recognition particle (SRP), 6
Signal sequences, 90
Simian virus 40 oncogene, 232
Single-stranded RNA (ssRNA),
   247, 251, 253–255, 257, 258,
   260, 265, 270
SM buffer, 132, 135, 136
Sodium azide, 11
Sodium dodecyl sulfate (SDS),
   11, 69, 107, 127, 133, 134,
   159, 160
Sodium pentobarbital, 269
Sodium thiocyanate, 68
Somatic-cell hybridization, 157
Somatostatin, 241
Southern blots, 135, 139, 140,
   142, 160, 186, 201
SP6, 253
Spermine, 5
Spinal cord, 24
Spodoptera frugiperda, 105
SRP (*see* Signal recognition particle)
SSC, 133, 139, 161
ssRNA (*see* Single-stranded RNA)
Stop-anchor sequences, 90
Striatum, 204
Subfamilies, 96
Subiculum, 257
Substance-K, 57, 105
Subtractive hybridization, 50, 79, 80
Sucrose gradients, 224
Superfamilies, 96
SV40, 194, 195, 206, 234
Synthetic oligonucleotide probes, 255

T3, 253
T4 DNA polymerase, 30
T4-infected *E. coli* (T4 DNA ligase), 17
T4 Polynucleotide kinase, 258
T7, 253
T7 RNA polymerases, 19, 257
T8 surface marker, 201
Tac promotor, 104
TATAA element, 146, 147
TBE, 42, 127, 129, 144, 146
TCA (*see* Trichloroacetic acid)
T-cell receptor, 57
TEA (*see* Triethanolamine)
TEMED, 42, 144
Terminal deoxynucleotidyl transferase (terminal transferase), 16, 258
TES, 39
Tetracycline, 18
Thr (*see* Threonine)
Three-dimensional (3-D) reconstruction, 265, 266
Threonine (Thr), 5, 90, 91
Thymidine kinase, 158, 232
Thymidylate, 15
Thyroglobulin, 99, 100
Thyroid C cells, 118
Transcription, 231, 232, 271, 276
Transfection, 18, 20, 183, 206
Transfer RNA (tRNA), 50, 146
Transferrin, 89, 118
Transformation, 18
Transgenic mice, 221, 226, 228–230, 233, 234
Transmembrane domains, 87–90
Trasylol, 107
Trichloroacetic acid (TCA), 8–10, 26–28, 67, 272
Triethanolamine (TEA), 6
Triton, 8, 250
tRNA (*see* Transfer RNA)
Trp (*see* Tryptophan)
Trypsin, 208
Tryptophan (Trp), 5, 55, 93
TTBS, 107, 108

Index

TTP, 67
Tunicamycin, 182
Tween-20, 108
Tyr (*see* Tyrosine)
Tyrosine (Tyr), 5, 90, 92
 hydroxylase, 94, 96
 kinases, 91, 205

UNC-phenol, 120, 129, 130, 137, 143, 146
Universal sequencing primer, 40
Untranslated regions, 85
Urea, 42, 144, 146
UTP, 254, 272

Val (*see* Valine)
Valine (Val), 5, 87
Vasopressin, 91, 262
Vectors, 14, 18–20, 23, 65, 97, 103–105, 120, 121, 126, 142, 144, 177, 180, 181, 188, 194, 208, 224, 225, 253

Viral envelope protein, 178
Virion, 182, 184

Western blots, 105, 106, 135
Wheat germ, 1–3, 9

Xenopus oocytes, 105
X-gal (*see* 5-Bromo-4-chloro-3-indolyl beta D-galactoside)
X-ray crystallography, 83
Xylene cyanol blue, 127, 128
Xylene cyanole, 40

Yeast artificial chromosomes, 121
YT broth, 38

Zinc fingers, 96
Z-value, 95